Promoting farm/non-farm linkages for rural development

CASE STUDIES FROM AFRICA AND LATIN AMERICA

Edited by
Benjamin Davis
Agriculture and Economic Development Analysis Division
FAO Economic and Social Department

Thomas Reardon
Department of Agricultural Economics
Michigan State University

Kostas Stamoulis
Agriculture and Economic Development Analysis Division
FAO Economic and Social Department

Paul Winters
Mexico, Central America and Hispaniola Region
Inter-American Development Bank

FOOD AND AGRICULTURE ORGANIZATION OF THE UNITED NATIONS
Rome, 2002

ISBN 92-5-104868-1

Foreword

Empirical evidence highlights the importance of off-farm activities in the income-generating portfolios of rural households in developing countries. It is critical to determine how such activities can be promoted, given the importance of non-farm income as a mechanism whereby rural households can sustain and improve their livelihoods and as a possible path out of poverty. Particular attention needs to be paid to ways in which spin-off activities in the non-farm sector can be promoted through policy and programme action in the presence of agricultural growth. Spin-off activities can emerge from backward and forward production linkages with agriculture, expenditure linkages that come with rising agricultural income, or investment linkages as non-farm income alleviates cash constraints faced by households.

In this publication these dynamic linkages and spin-off activities are explored in a series of case studies in Africa and Latin America. The objectives are: (i) to characterize the spin-off activities in each study area and evaluate their importance to rural employment, incomes and growth; (ii) to describe, compare, analyze and synthesize experiences — successful and unsuccessful — of growth and promotion of linkages in high potential areas; and (iii) to devise policy and programme options that would interest policy-makers looking to achieve agricultural growth in high-potential areas and promote growth and employment opportunities in the off-farm sector in rural economies.

The case studies focus particularly on the institutional, organizational and technological aspects of spin-off activities. This include the rules of the game such as contracts and standards governing economic relations, the players such as associations and intermediaries and the instruments that can be used to translate agricultural sector growth into activities up and down-stream. In order to give adequate consideration to the importance of these aspects under different settings, the case studies include low-income countries such as Ghana and Ethiopia as well as middle-income countries such Peru and Mexico. The challenges in the low-income countries include limited local demand and limited investment funds for spin-off activities. In middle-income countries, the challenge is for spin-off activities to stay local, because there are often incentives for farmers to buy inputs from non-local sources and for processing to take place outside local areas.

The case studies indicate in general that the public sector and non-governmental organizations (NGOs) play an important facilitating role to private initiative in developing linkages between agro-industry and farmers. This role may include organizing farmers or assisting NGOs or private enterprises to take on responsibilities previously discharged by states, providing credit, facilitating access to inputs, providing information on technology and ensuring that contract requirements are met.

Since the present case studies were initiated, interest in overall rural development issues has greatly increased. Rural strategies by development institutions (including development banks) now give a prominent position to the role of agriculture in increasing rural incomes and rural employment through its impact on activities up and down-stream of agricultural production. In publishing this volume, it is our hope to contribute to this debate with more concrete examples and provide incentives for more such undertakings.

Rome, November 2002

Prabhu Pingali
Director, ESA

Contents

Acknowledgements

The authors would like to thank Jacques Vercueil for his constant encouragement and most helpful comments and suggestions through a large part of the research. We would also like to thank Marina Pelaghias for putting together the printed version of this paper and Joanne Morgante for the cover design. The content is exclusively the responsibility of the authors and does not necessarily reflect the position of FAO or its governing bodies.

List of acronyms

ADLI	Agricultural development-led industrialization
ASIA	Asesoría y Servicios Integrados Agropecuarios
CGE	Computable general equilibrium
COSCA	Collaborative Study of Cassava in Africa
CSA	Central Statistics Authority
FIRA	Fideicomisos Instituidos en Relación con la Agricultura
FOB	Free on board
GAMS	General algebraic modelling system
GATT	General Agreement on Tariffs and Trade
GDP	Gross domestic product
GLSS	Ghana Living Standards Surveys
GPRTU	Ghana Private Road Transport Union
GVO	Gross village output
GVP	Gross village product
HASIDA	Handicraft and Small Industrial Development Agency
HCDA	Horticultural Crops Development Authority
IFAD	International Fund for Agricultural Development
INEGI	Instituto Nacional de Estadística Geografía e Informática
I-O	Input-output
ITTB	Tigray Regional State Bureau of Industry, Trade and Transport
ITTU	Intermediate Technology Transfer Unit
LDC	Least developed country
LSMS	Living Standard Measurement Study

NGO	Non-governmental organization
PEAT	Programa de Asistencia Técnica para Apoyar la Producción de Granos Básicos
PROCREA	Credit for Administration Programme
REST	Relief Society of Tigray
RNFE	Rural non-farm economy
RTPD	Rural Technology Promotion Department
SAM	Social accounting matrix
TCG	Trade and Commodity Group
TDA	Tigray Development Agency
VCGE	Village computable general equilibrium

Chapter 1
Promoting farm/non-farm linkages in developing countries

Benjamin Davis, Thomas Reardon, Kostas Stamoulis and Paul Winters

INTRODUCTION

Evidence from developing countries points towards the growing importance of non-farm activities in the income-generating portfolio of rural households (Lanjouw and Stern, 1993; Estudillo and Otsuka, 1998). From an extensive review of the literature, Reardon *et al.* (1998) show that rural non-farm activities account for 42 percent of the income of rural households in Africa, 40 percent in Latin America and 32 percent in Asia. It is critical to determine how such activities can be promoted, given the importance of non-farm income as a mechanism whereby rural households can maintain their livelihoods and as a possible path out of poverty. Particular attention should be paid to ways in which spin-off activities in the non-farm sector can be promoted in the presence of agricultural growth. Spin-off activities can emerge from backward and forward production linkages with agriculture, or through expenditure linkages that come with rising agricultural income.

In this volume, several case studies of farm/non-farm linkages are presented in which spin-off activities already exist. The purpose of the case studies is to explore the ways in which spin-off activities were promoted and to consider how the activities might be further supported. The case studies focus particularly on the following aspects:

- institutional: rules of the game such as contracts and standards that govern economic relations;

- organizational: players such as associations and intermediaries;

- technological: instruments that can be used to promote spin-off activities from the agricultural sector.

To give adequate consideration to the importance of these instruments, attention was paid to the differences in promoting spin-offs in low-income countries such as Ghana and Ethiopia as opposed to middle-income countries such Peru and Mexico. The challenges in the former group include limited local demand and limited investment funds for spin-off activities; the challenges in the latter group are for spin-off activities to stay local, because there are often incentives for farmers to buy inputs from non-local sources and for processing to take place outside local areas.

The purposes of this introductory chapter are to lay the groundwork for the case studies and to summarize the results. The chapter is set out as follows: Section 2 gives a brief conceptual framework, explaining the types of spin-off activities that are linked to agriculture, Section 3 provides an overview of the six case studies presented in this volume and Section 4 notes the key points from the case studies concerning actions taken to spur spin-offs in terms of institutions, organizations and technologies.

CONCEPTUAL FRAMEWORK

The literature identifies two major types of farm/non-farm linkages: production and expenditure. Production linkages can be further divided into backward and forward linkages, or, to use an alternative terminology, up-stream and down-stream linkages. Backward production linkages refer to linkages from the farm to the part of the non-farm sector that provides inputs for agricultural production, for example agrochemicals. Forward production linkages refer to the part of the non-farm sector that uses agricultural output as an input. The distribution and processing of agricultural outputs are fundamental components of forward production linkages.

Expenditure linkages refer to the fact that households deriving income from one type of activity, farm or non-farm, are likely to spend that income on products of other activities. Farmers buy non-farm products with income generated from agriculture. Local entrepreneurs and wage earners use income from the sale of non-farm products to buy food and other agricultural outputs. Expenditure linkages can be divided into consumption and investment linkages. Consumption linkages refer to expenditures related to household consumption; investment linkages refer to expenditure used to finance farm or non-farm activities. Investment linkages can be particularly important within households. Returns on farm activities may be invested to initiate or expand non-farm activities and vice versa.

Different types of spin-off activities will emerge, depending on the structure of the agricultural sector and the type of growth that is occurring. If agriculture requires significant external inputs, growth in backward production linkage activities are to be expected. If output from agriculture requires processing before it can be sold, or if there is significant value added by processing, forward production linkages are to be expected. If there is sufficient growth in the agricultural sector to induce rural income growth, expenditure linkages will induce growth in consumption and possibly investment. Dynamic agricultural sectors are more likely to have multiple and diverse linkages. Growth in horticultural production, for example, with output exported or sold to urban markets, is an extreme case where substantial input requirements, high-value output and significant cash income are likely to create numerous local linkages. At the other extreme, staple products are less likely to create local linkages because they provide little cash income and tend to be low-input and consumed without processing. An exception is cassava, discussed in Chapter 7.

The non-farm economy that emerges or fails to emerge in the presence of agricultural growth is conditional on incentives to potential investors and capacity to undertake such activities. Incentives are largely driven by the profitability of an activity, which will depend among other things on the macroeconomic framework, output and input prices and the risk associated with the activity. The capacity to invest in non-farm activities will be determined by the vector of assets – human, physical, financial, social and public – owned by the individual, household or community. Incentives and capacity to invest are strongly influenced by institutions (rules of the game) and organizations (players in the game) and by the technologies available. This means that the state and civil society play an important role in determining how the non-farm economy responds to agricultural growth.

The direction taken by the non-farm economy depends on local conditions, even with appropriate incentives and a degree of investment capacity. Spin-off activities may not stay local if farmers purchase inputs from distant sources, if output is processed in remote locations and if farmers use income from goods imported into the region. The ability of a locality to capture the benefits of the non-farm economy depends on local incentives and capacity, which are directly influenced by government policy. As agriculture is modernized, the concept of "local" is bound to become larger because of economies of agglomeration in local intermediate cities and metropolitan areas. How the expansion occurs and how great it becomes can be influenced by government policy.

The types of linkages are explored in each of the case studies in this volume. Attention is given to incentives, to capacity to invest in non-farm activities and to the influence of local conditions.

CASE STUDIES

Six study regions, three in Latin America and three in sub-Saharan Africa, were identified as being suitable cases for examining non-farm employment spin-offs. Studies from these regions form the basis of each chapter. An overview of each study is given here.

Chapter 2 (J. Edward Taylor and Antonio Yunez-Naude) examines farm/ non-farm linkages in Mexico using two methodologies: a village/town social accounting matrix (SAM) and a village/town computable general equilibrium (CGE). These models allow detailed examination and understanding of linkages between the farm and non-farm sectors and calculation of multiplier effects. The results indicate that although demand linkages are important, the largest of these is with markets outside the local economy: a large share of inputs, consumption and investment goods purchased by rural households is supplied by regional urban centres. Village households are diversified away from agriculture, mainly through family participation in labour markets outside villages, through wage work or through migration to distant urban centres or abroad. They find that where technological and other constraints limit the supply responsiveness of agriculture, measures must be enacted to improve supply response if farm/non-farm linkages are to be strengthened.

Chapter 3 (Fernando Rello and Marcel Morales) also focuses on Mexico, in particular on the state of Querétaro. This state is chosen because of the dynamic nature of the agricultural sector, which is largely a product of the proximity of Mexico City and Querétaro's strong links with the hinterland. The chapter presents several case studies on agro-industrial systems, which embody the characteristics and development of the systems. The chapter examines the geographic linkages between agriculture and small and medium-sized towns and intermediate cities. One of the main points made by the paper is the importance of agro-industries in forging links, quite often in conjunction with public agencies and non-governmental organizations (NGOs). NGOs and certain public agencies have become particularly important because of the institutional vacuum created by the withdrawal of the state in a number of areas in Mexico. An important and expanding institutional development noted in the case studies is the use of contract farming for certain commodities. NGOs and public agencies have helped to facilitate these types of relationships. Finally, the paper notes that links between the farm and non-farm sectors depend on the scale of transactions, with smaller purchases such as seed, animal feed and repairs made from small towns, mid-level products such as fertilizers and agrochemicals from medium-sized towns and large purchases such as tractors and trucks from intermediate cities.

Chapter 4 (Javier Escobal and Victor Agreda) examines recent institutional innovations in two regions of Peru that have altered the relationship between farmers and agro-industrial firms. These institutional innovations included contract farming and share contracts in which managerial services were traded for labour services and land. The results indicate that the innovations were successful in improving the quality of farm/non-farm linkages. The success of these innovations has been partially the result of a combination of public goods and services and sufficient private assets, including managerial ability. In the case of asparagus, however, contracts tended to favour large producers at the expense of small producers. In the case of cotton, on the other hand, the emergence of farmer companies increased employment of smallholders and their incomes. The results indicate that the benefits of institutional innovation depend on several factors, including the crop characteristics, characteristics of farmers, public goods and services available and NGO involvement.

Chapter 5 (Tassew Woldenhanna) focuses on farm/non-farm linkages in the marginal Tigray region of northern Ethiopia. The study is different from other studies in Africa, which have tended to focus on dynamic regions. The basis of the study is survey data collected in two regions of Tigray and on secondary data collected from national and regional government offices. The results indicate that backward and forward production linkages are limited and that expenditure and specifically consumption linkages are the strongest form of linkage, as is the case elsewhere in Africa. The analysis found, for example, that 86 percent of total expenditure is on regionally produced farm and non-farm output, and that although non-farm expenditure remains small at 21 percent of the total, it increases in importance as income increases. The results show the importance of income diversification and agricultural productivity and that farmers with off-farm income tend to be more productive. The chapter concludes with a number of policy implications, including the need for institutional support for developing linkages, the importance of rural towns and targeting of specific vulnerable groups for inclusion in the benefits of expanding non-farm activities.

Chapter 6 (Lydia Neema Kimenye) provides a detailed case study of the French bean processing industry in Kenya. The French bean was chosen for the case study because the Kenyan government has singled out horticulture as a high-potential growth area; the sector has been expanding in recent years through the export and frozen-vegetable markets. One of the main features of this market is the dominance of contract farming as the primary means of interaction between farmers and agro-industry. As part of these contracts, farmers tend to get inputs from the processing firms and as a result production linkages in the local region tend to be limited, with output going to the processing firms and inputs purchased in bulk by the processing firms from urban suppliers. The benefits of French bean processing to the local economy tend to come primarily from the income gains of contracted farmers in the region and their expenditures in local markets. Direct employment linkages to processing firms, although not substantial, are potentially helpful to the local economy through expenditures. One of the interesting results of the case study is that one of the processing firms lost a market outlet as a result of quality problems and did not pay its contracted farmers. Although contracting may appear to be a safe market for farmers, it is clearly risky; the negative effects on a local economy can be substantial.

Chapter 7 (Ramatu Al-Hassan and Irene Egyir) examines the cassava subsector in Ghana. Cassava is almost always sold in a processed form, so it has a high potential for non-farm linkages. Ghana has recently expanded its export market for cassava chips, enhancing its value in the market. The focus of the study is two high-potential agricultural regions that produce cassava chips and an alternative local processed cassava product – *kokonte* in Atebubu and *gari* in Nkwanta. Because it is a low-input agricultural product, there are limited backward production linkages, but because processing is necessary for the market and because of high transportation requirements, there are substantial forward linkages. One of the benefits of the expansion of the chip market in recent years has been the rise in the price of processed cassava, including locally processed kokonte and gari. Weaknesses in the export market for chips, however, partially for domestic reasons, could limit this market and adversely affect cassava production and linked industries. The study highlights the effects of market changes and the importance of policies to complement private initiatives.

SYNTHESIS AND POLICY IMPLICATIONS

The case studies in this volume offer insights into farm/non-farm linkages and suggest actions that might be taken to promote them. In this section, the findings of the case studies are synthesized and implications are drawn for policy and programme implementation. These implications are divided into the institutional, organizational and technological instruments that can be used.

Institutional

Contract farming is one mechanism that can help to overcome market and organizational failures and link farmers with agribusiness; it has the potential to provide substantial benefits to farmers, producers and the rural economy. In order to provide these benefits, the state may in some circumstances act as a facilitator, or third party, in triangular contracts. This facilitation might come in the form of credit or technical assistance. An alternative to state involvement is involvement of NGOs or even private entities as third parties. One of the downsides of contract farming is the tendency of agro-industry to purchase inputs from outside the production region, thus limiting local backward linkages. Although contracts may appear to be a low-risk alternative to selling on the spot market, there is a risk that contracting firms may face difficulties and fail to honour contracts. States should assist in developing this type of relationship and must ensure, through legislation and their judicial system, that the rights of contracted farmers are protected.

Evidence from the case studies suggests that entry barriers may limit the ability of some households to participate in non-farm activities. This may exacerbate income inequality, because wealthier households are able to enter into lucrative non-farm activities and expand income, while poorer households remain in low-return non-farm activities. A particular problem is lack of access to credit, which can limit linkages in a number of ways. It may limit the ability of households to enter into non-farm activities or expand their current activities, and may limit farmers' ability to take advantage of opportunities for selling to agribusiness. Access to credit will not guarantee expansion of non-farm activities, but credit limitations can hinder development of such activities, because credit is often necessary for entry into and expansion of non-farm activities. Credit may also be necessary for new crops or new technologies that must be adopted to produce quality output for processing. If non-farm activities are to develop, states need to assist with credit access when markets do not function well.

Infrastructure and location have a substantial influence on the creation of linkages; they can also be a barrier to entry, because poor infrastructure limits opportunities. Governments need to invest in infrastructure such as roads, electricity and telecommunications and other public goods, primarily human capital such as education and health, that foster the development of non-farm activities and increase their productivity. Rural areas close to urban centers tend to have greater farm/non-farm linkages. Rural towns play a significant role in agricultural development through linkages between the non-farm and farm sectors; investment in infrastructures to promote non-farm activities should concentrate on these locations.

Technological

Backward production linkages in Africa are generally limited because of the low inputs in agriculture. Forward production linkages depend largely on the commodity being produced and the type of processing. Expenditure linkages are critical for African rural development: they are the primary mechanism by which agricultural growth affects the non-farm sector. To foster the development and expansion of farm/non-farm linkages in Africa, there must be an emphasis on improving agricultural technology. Backward and forward production linkages require modern agricultural-production systems. Governments must consider actions that simultaneously promote complementary non-farm activities such as input supply and output processing as well as promoting agricultural technologies.

In Latin America, there are more backward and forward production linkages; expenditure linkages within local economies are minimal, because farmers tend to purchase items produced in distant urban centres. As agriculture develops, the backward and forward production linkages tend to become less local in that there is a tendency for many non-farm activities to move to regional centres. This is often a result of modernization of the agricultural sector, because transaction costs are inevitably reduced, allowing such activities to shift from small local producers to larger regional producers. This is not necessarily a problem, but regions should be assisted in developing regional centres that can foster farm/non-farm linkages.

Organizational

In general, agribusiness plays an important and proactive role in creating linkages to agriculture. Farmers often play a less important role in forging linkages although without their active participation the linkages will not work. Given that agribusiness has taken a lead in developing linkages between the farm and non-farm sectors, it is important that the state create an environment that is conducive to investment in agribusiness activities. One reason for the lack of farmer initiation in developing farm/non-farm linkages is that farmers tend not to be organized sufficiently to initiate new activities. The lack of significant producer organizations limits the ability of farmers to be proactive in forming linkages. Taking actions to promote these organizations is likely to lead to a more dynamic agricultural system.

CONCLUSION

The numerous structural-adjustment and stabilization programmes that have been implemented in developing countries have resulted in states withdrawing to some extent from agriculture and rural areas. This has created an institutional vacuum that has not been adequately filled. States cannot and should not play the role they once did in agriculture, but they still need to be engaged. States must reconsider their role in rural development and ways in which they can foster and expand linkages. The case studies indicate that the public sector and NGOs as well as private entrepreneurs play an important facilitating role in developing linkages between agro-industry and farmers. This role may include organizing farmers or assisting NGOs or private enterprises to take on responsibilities previously discharged by states, providing credit, assisting with inputs, providing information on technology and ensuring that contract requirements are met. In this way, the public sector, NGOs, and private entrepreneurs are helping directly to create beneficial linkages between agro-industry and farmers, and indirectly creating other linkages between the farm and non-farm sectors.

REFERENCES

Estudillo, J.P. & Otsuka, K. 1999. Green revolution, human capital and off-farm employment: changing sources of income among farm households in central Luzon, 1966–1994. *Economic Development and Cultural Change*, 47(3): 497–523.

Lanjouw, P. & Stern, N. 1993. Markets, opportunities and changes in inequality in Palanpur, 1957–1984. *In* A. Braverman, K. Hoff & J. Stigliz, eds., *The economics of rural organization: theory, practice and policy*. New York, USA, OUP.

Reardon, T., Stamoulis, K., Cruz, M.E., Balisacan, A., Berdegue, J. & Banks, B. 1998. Rural non-farm income in developing countries. *In FAO: the state of food and agriculture, 1998*. Rome. (FAO Agriculture Series, No. 31.)

Chapter 2
Farm/non-farm linkages and agricultural supply response in Mexico: a village-wide modelling perspective

J. Edward Taylor and Antonio Yunez-Naude

INTRODUCTION[1]

Rural farm/non-farm linkages are critical in shaping the impacts of policy and market changes on rural economies. In towns and villages, commodity and factor markets transmit the impacts of exogenous shocks from directly affected households to others in the local economy, creating local income multipliers. Market linkages between towns and villages then diffuse local impacts into larger economies such as regional ones, setting in motion income multipliers and local general-equilibrium feedbacks there, and possibly generating feedbacks to the village or town originally affected by the income shock. Understanding the nature of farm/non-farm growth linkages is a first step in designing policies and programmes to exploit these linkages and promote broad-based economic growth in least-developed countries (LDCs), and rural industrialization.

The range of activities and income sources embodied in the rural non-farm economy (RNFE) depends on definitions of "rural" and "urban". One extreme is to consider urban areas as major cities: in this case, the urbanization of rural areas associated with industrialization is regarded as rural industrialization rather than change from rural to urban areas (Otsuka and Reardon, 1998). In other definitions, urban areas include small and medium-sized cities. The RNFE comprises a wide range of activities and sources of income, from local rural non-farm activities and local labour markets to involvement of villages or towns in regional, national and international markets for inputs, products and labour.

The extent to which linkages transmit the impacts of changes in rural household incomes across sectors is an empirical question. Two extremes are possible. At one extreme, households, villages or towns directly affected by these changes may act as enclaves, supplying few inputs and demanding few

---

[1]   The authors thank George Dyer, Javier Becerril, Virginia Evangelista, Xochitl Juarez and Mara Ruiz for their superb research assistance.

goods from others in the local or regional economy. This is the case of many villages in LDCs, where exogenous income changes such as income transfers, including migrant remittances, influence production, income and demand in the households concerned, but create few farm or non-farm income linkages within the village. In this case, a regional analysis is required to uncover rural income multipliers.

At the other extreme, the directly affected households may be closely integrated with local product and factor markets, supplying inputs to local production and demanding locally produced non-tradables. In this case, exogenous policy or market shocks may stimulate incomes in the directly affected households and in others in the local economy. If expenditure linkages favour non-farm goods, and if the supply response in non-farm production is elastic, transmission of policy and market shocks through local economies and across sectors may be substantial. The resulting farm/non-farm income linkages may be manifested primarily in a local economy if there is an important local non-farm sector producing non-tradables, or primarily between local and regional or national economies if village income changes stimulate demand for non-farm goods in urban centres.

In either case, resource and technological constraints may be instrumental in determining the magnitude and direction of the impacts of exogenous policy or market changes on village and town economies. If villages are defined as conglomerations of fewer than 10 000 habitants, manufacturing activities are generally absent in villages in Mexico, as in most LDCs. Village populations, however, carry out local non-farm activities and have rural non-farm incomes; their economies are linked with the outside world through regional product and labour markets.

This chapter discusses the ways in which linkages between the farm and the non-farm sectors at the village level can be studied; the second section proposes two methodologies for doing so. First, a village-town social accounting matrix, or SAM, is applied to five villages of medium and small farmers and to a "mini-region" or municipality in rural Mexico. The third section describes the village and town data. Findings from village and village/town SAM multiplier models are reported in the fourth section. In the fifth section, the second methodology, a village/town CGE model, is estimated and used to examine the impacts of exogenous income changes on the economy of the mini-region. Discussion of policy implications of the results and issues for future research conclude the chapter.

FOUR MODELS OF FARM/NON-FARM LINKAGES

Three approaches have dominated development economics research on farm/ non-farm linkages: the input-output (I-O) approach, the expenditure system approach and village-wide modelling. CGE modelling techniques have recently been developed to study village and small regional economies, offering a new perspective on the impacts of exogenous income changes on rural economies.

Input/output models

I-O or Leontief models offer a snapshot of linkages across production sectors in national, regional or village economies. The elements in the rows of an I-O matrix represent sales of output from a row sector to other sectors, shown in the columns; that is, forward linkages. The column elements represent backward linkages, purchases of inputs by column sectors from other row sectors. In general, the larger the elements in the row and column for a sector, the larger the sector's potential to stimulate growth through creation of forward and backward linkages. The extreme case of all zero entries in a Leontief matrix corresponds to an economic enclave devoid of linkages. Leontief multipliers calculated from I-O matrices measure the multiplicative effect of changes in final demand for sectoral outputs when the household sector, investment, government and rest-of-the-world demands are treated as exogenous.

The first studies of farm/non-farm growth linkages were based primarily on input-output models. Hirschman (1958), citing the sparseness of rows and columns corresponding to traditional agriculture in I-O matrices for countries, criticized agriculture for its lack of forward and backward linkages with the rest of the economy. This view was challenged, most notably by Adelman (1984) and Mellor (1976), and the list of potential linkages has been expanded from production to consumption and fiscal linkages. Even if staple production generates few backward and forward production linkages, for example, a change in exogenous demand for staples, such as export, may raise the incomes of staple-producing households. These households may in turn spend their new-found income on goods and services that include agricultural and non-agricultural commodities. Their expenditure demands may thus stimulate a new round of production increases as firms expand their output to satisfy household consumption demands and increasing demand for intermediate inputs. If staple-producing households have expenditure patterns favouring non-agricultural goods, the increased demand for staples may generate important production linkages through household expenditures.

Expenditure-system models

The potential importance of rural households as a source of demand for farm and non-farm goods stimulated the growth of a new literature on farm/non-farm linkages focusing on rural-household expenditure patterns. Mellor (1976) emphasized the importance of consumption linkages in transmitting changes in rural incomes into demand for farm and non-farm goods, tradables and non-tradables. Utilizing an expenditure system approach, Mellor (1976), Hazell and Roell (1983), Rangarajan (1982) and others offer compelling evidence for the existence of farm/non-farm linkages.

The approach used in this "new economics of growth" research generally entails estimating household expenditure functions for various classes of goods – farm and non-farm, tradables and non-tradables. The estimated equations are used to ascertain farm/non-farm linkages by comparing the impacts of changing farm incomes, for example those resulting from a new agricultural technology, on demand for these goods. Hazell and Roell (1983), for example, concluded that in Muda (Malaysia) and Gusau (Nigeria) the share of increments in total household expenditures allocated to foodgrains is lowest in high-income households, and that the share allocated to local non-tradables is highest in these households.

The expenditure system approach has two important limitations. First, it is partial. Estimated marginal propensities to spend income on farm and non-farm goods are used to identify and quantify farm/non-farm linkages. Modelling expenditures by different household groups is a critical first step in modelling potential farm/non-farm linkages. Without data on the production side, however, it is not possible to ascertain whether these potential linkages actually stimulate growth in the local economy, because the local content of locally supplied goods is not known; nor is the local content of household investment demand known. These shortcomings are the basis for Hart's (1989) critique of Hazel and Roell (1983). Hart argues that Hazel and Roell's predictions of growth linkages were exaggerated, precisely because they did not take into account these production and investment considerations.

A second limitation of most of the growth-linkage literature is that it generally assumes perfectly elastic supply responses, ruling out price effects in what are essentially Keynesian demand-driven models. This assumption may be valid under some circumstances, such as when under-utilization of local factors makes the supply of farm and non-farm goods highly responsive to changes in demand, with little or no inflationary pressure. Where non-linearities in production and local resource constraints create less than perfectly elastic supply responses, however, fixed-price linear models are likely to exaggerate income and production effects of policy changes (see Taylor 1995; Haggblade, Hammer and Hazell, 1991). CGE models incorporating non-linearities and endogenous prices have been utilized extensively at the national level. These models generally treat rural households as homogeneous, however, and do not take into account the diversification of rural economies, through which many of the most important farm/non-farm linkages are likely to play themselves out. The diversification of rural economies from farm to non-farm activities is one of the most pervasive and far-reaching features of contemporary LDCs. As rural economies become increasingly diverse, farm/non-farm linkages take new forms, with impacts felt increasingly in small and intermediate cities emerging in traditionally agricultural areas instead of in traditional urban centres. There also appears to be a diversification away from traditional crop activities in agricultural households (see Reardon, Delgado and Matlon, 1992; Taylor and Adelman, 1996).

SAM models

SAM models are designed to capture the complex interlinkages among production, institutions – including households – and the outside world. SAMs provide a starting point for village economy-wide analysis. They are a useful expository device for portraying the structure of rural economies, and they are the basic data input for SAM multiplier and general-equilibrium modelling (Taylor and Adelman, 1996). They summarize and neatly illustrate flows of inputs, output and income between food production and other productive sectors in an economy, flows of income between production activities and households, channelling of household incomes into consumption and investments and exchange of goods and factors between an economy and the rest of the world. The great strengths of the SAM are its comprehensiveness and flexibility in adapting to diverse institutional settings and economic structures and in providing an accounting framework to address diverse policy and planning issues.

Village SAM models

In theory, a SAM can be developed for any economy, from the world to a village to an individual household. The first application of SAMs to village economic analysis appears in Adelman, Taylor and Vogel (1988). Since then a number of other village SAM studies have been carried out (see for example Lewis and Thorbecke, 1992; Subramanian and Sadoulet, 1990; Parikh and Thorbecke, 1994). The village social accounting framework, like its national counterpart, is a form of double-entry accounting. It presents accounting entries in income and product accounts and in input-output production accounts as debit and credit entries in income balance sheets of institutions and activities. Activities may include farm and non-farm production, or any disaggregation of the two. Institutions in SAMs typically include different household groups, governments and the rest of the world. In the context of economy-wide modelling, institutions are categories of economic actors. It is assumed, of course, that all members of a given category of actors interact in a similar manner with other categories and activities in a village.

Entries in a SAM include:

- intermediate input demands between production sectors (the Leontief matrix is contained within the SAM);

- income or value-added paid by production sectors to different types of labour such as men or women, educated or non-educated or different ethnic groups, land or capital;

- distribution of labour or capital income across different household groups;

- distribution of household group expenditures across savings, consumption of domestically produced goods and services, and imports.

A government account collects income from activities and households and redirects it within the system to government demand for goods and services, transfers to production activities or household groups, saves it, or uses it to pay foreign institutions for imported goods and services or to repay debts.

The total product of each activity must be allocated to some use inside or outside the economy, such as intermediate demand, consumption, investment, government demand or exports. Gross receipts of each activity must be allocated to some entity inside or outside the system, including purchases of inputs from other activities, payments to labour and capital, imports, taxes and savings. A salient characteristic of SAMs, derived from double-entry accounting, is that equality must be maintained between the sum of expenditures – the column total – and the sum of revenues – the row total – for every account in the system.

Village SAMs have the same conceptual framework as national SAMs, but they depart from national SAMs in specific ways related to the unique nature of village economies and institutions. Some examples of village production activities include subsistence food production, wood gathering, export crop production, handicrafts, traditional healing and religious activity in a temple or mosque. Examples of village institutions are households grouped by land holdings or base-period calorie intake, compounds grouped by compound landholdings, internal and international migrants and village schools. The rest of the world in village models includes the regional and national economies outside the village and the world economy, which does not share the same currency as the village. Examples include marketing cooperatives, plantations, weekly markets, nearby centres or towns, labour commuting and domestic and foreign migrant labour markets. For a village, the rest of the world includes regional and national governments.

One conceptual departure of the village SAM from the national SAM is that rest-of-the-world subaccounts of village SAMs do not necessarily balance. A regional or national government, for example, may be a net surplus appropriator or a net subsidizer of a village. Remittances to village households from migrants abroad through a foreign account may not necessarily be used to purchase goods from abroad, but may be used to purchase goods produced in the village or brought into the village from regional or national markets. Methodologically, these inconsistencies are addressed through the use of entries representing payments between rest-of-the-world accounts or through aggregation, for example by combining some rest-of-the-world accounts, the sum of whose transactions with the village must balance. Another departure of village SAMs from national SAMs is that in village SAMs, non-monetary transactions are typically important. These non-monetary transactions include production for own consumption, labour exchanges or labour lending, interlinked factor markets, interlinked factor and non-factor-input markets and access to commons.

SAM multiplier models

To move from the village SAM as an accounting framework to a village model first requires assumptions about the behaviour of village actors and the specification of production functions. The SAM summarizes transaction flows among economic actors in the village. In designing village models, the simplest assumption is that the responses of village actors to income changes are strictly proportional to the total level of activity in each account – that is, the column totals in the SAM. This means that on the expenditure side, marginal expenditures by village institutions equal average shares derived from the SAM; on the production side there is a fixed input-output technology. These assumptions are restrictive but necessary to estimate fixed-price village SAM multipliers (described below), which are analogous to the Leontief multiplier in input-output analysis. These multipliers are the basis for the SAM policy experiments in Adelman, Taylor and Vogel (1988), Golan (1990), Subramanian and Sadoulet (1990), Lewis and Thorbecke (1992), Ralston (1992) and Parikh and Thorbecke (1996).

Constructing a village model also requires specifying which accounts in the village SAM are endogenous and which are exogenous. This choice is critical in modelling the impact of change on village economies, because the modeler is strictly speaking free to change only exogenous variables and model parameters. The endogenous accounts in the model capture the responses of village economic actors to changes in the exogenous accounts or in parameters. In village models, the logical choices for exogenous accounts are the government and the rest of the world. If the village capital market is fully integrated with outside capital markets, it may be treated as exogenous as well. In most LDC rural areas, however, capital markets are local at best, with the result that local savings constrain investments. In this case, capital markets should be included in the endogenous SAM accounts. Capital is treated as endogenous in all five of our village models for Mexico.

The village multiplier matrix contains estimated total direct and indirect effects of exogenous income injections on the endogenous accounts in the village SAM. The village Leontief multiplier is one component of the village SAM multiplier. The SAM multiplier, however, also captures expenditure linkages induced by changes in production activities through their effect on institutional incomes in the village. These expenditure linkages are typically stronger than production linkages in village SAM models.

Linkages between production and factors, between factors and households and between households and production shape the impact of exogenous changes on a village economy. The village multiplier consists of multiple rounds of feedback among subaccounts in the village SAM. Each new injection of income into a SAM subaccount first swirls around the local subsystem of accounts and is then transmitted to other subsystems of the SAM. This process continues as part of the new income generates a derived demand for goods and services or induces a redistribution of income flows within the village, while some leaks out.

The major strength of SAM multiplier models for estimating farm/non-farm linkages is that they integrate the I-O and expenditure-system approaches into a single model that captures production linkages, consumption linkages and interactions between the two. They thus offer a much more comprehensive and potentially reliable means to identify and estimate the importance of farm/non-farm linkages. Both the I-O and expenditure system approaches may be viewed as representing special cases of SAMs. Leontief multipliers are one component of SAM multipliers, because they ignore consumption linkages operating through the household sector. Multipliers obtained from expenditure-system estimates may be viewed as another component of SAM multipliers, in which production-side linkages are ignored.

Despite its strengths, the village SAM multiplier has the same basic limitations as its national counterpart, although some of these are less important in the village context. First, it is a fixed-price model. In a perfect neoclassical village economy, all transactions are for goods and services whose prices are determined by markets outside the village; but because the neoclassical village is assumed to be a completely open economy, the SAM village income multiplier of an exogenous income change is always unitary. An actual village is characterized by market imperfections that may cause village prices to diverge from market prices outside the village. In simulations using village SAMs, the critical questions are whether prices vary in response to exogenous changes and whether variations in price induce changes in the SAM share matrix. In general, the fewer the local resource and technological constraints on production, the stronger the case for using SAM multiplier models.

A second limitation of SAM multiplier models, related to the first, is the assumption that supply is perfectly elastic; that is, SAM models assume a Keynesian, demand-driven system. Even if supply response is elastic in the long term, it may in some cases be inelastic in the short term; crop production, especially for perennials, is a case in point. One way to incorporate inelastic supply response into SAM multiplier models is to impose constraints on production in the form of a perfectly inelastic supply response in some sectors, as in the second and third experiments described below; see Subramanian and Sadoulet (1990) and Lewis and Thorbecke (1992), or beyond predetermined output levels (Parikh and Thorbecke, 1994).

Third, SAM multiplier models assume that production utilizes linear, fixed-proportion technologies and that average and marginal expenditure propensities are the same. The second assumption can be relaxed by incorporating marginal rather than average shares into the SAM expenditure shares matrix prior to calculating the multiplier matrix. These marginal budget shares may be obtained from econometric expenditure-system estimates.

Village CGEs

Village CGE (VCGE) models overcome these limitations. They combine the strengths of microeconomic household-farm models (see Barnum and Squire, 1979; Singh, Squire and Strauss, 1986) with those of SAM-based, village-wide models. Although both SAM and VCGE models use village or mini-regional

SAMs as their data base, VCGEs capture price effects, nonlinearities in household-farm responses to policy changes and the implications of family resource constraints on production elasticities. Household-farm models are useful for estimating the effects of policy changes on production and consumption in individual households. They do not capture production and expenditure linkages among households, however, and recursive household-farm models ignore market imperfections, which can play an important role in household-farm economies (see de Janvry, Fafchamps and Sadoulet, 1991). These include factor-market imperfections caused by the high cost of separating management from labour on household farms (see Lopez, 1986; Bardhan, 1988). Household-farm micromodels also ignore general-equilibrium feedback effects (see Braverman and Hammer, 1986). As demonstrated above, village SAM models reveal income linkages among households that play an important role in shaping the impacts of policies on village incomes. In the fourth section of this chapter a microregion CGE is utilized to explore the impacts of exogenous income changes on the village-town microregion – that is replicating with a CGE the experiments in the third section of the chapter that use a SAM multiplier approach.

Structure of the village/town CGE model

The microregional CGE is centred on a household farm of the type prevalent in rural Mexico, engaged in maize production and a portfolio of other economic activities, including migration. The model captures production and expenditure linkages within villages, between villages and a nearby town and between the village/town microregion and the rest of Mexico, including household consumption and production demand for manufactured goods. Migration from Mexico to the United States and internal migration are modelled explicitly as a function of the returns to migration and the returns to family labour in the village/town microregion. The model consists of five blocks of equations for each of the village and town economies, linked by commodity and factor markets. The five equation blocks are a household-farm production block, a household-farm income block, an expenditure block, a set of general equilibrium closure equations and a price block.

In the town and each of the villages, household-farm production includes four production activities and one commercial sector that serves to import goods into the village from the town or the rest of Mexico or into the town from the villages or the rest of Mexico. Production in each of the sectors is carried out using two variable factors – family labour and hired labour – and two fixed factors – physical capital and land. In contrast to the traditional neoclassical household-farm model, it is not assumed that family and hired labour are perfect substitutes.

Household-farms are assumed to maximize utility defined by consumption goods and leisure. On the production side, this implies maximizing net farm income from the five production activities, given market prices for output and either market or shadow prices for factors of production and intermediate inputs. Endogenous shadow prices include family wages, which equal the marginal utility of leisure divided by the marginal utility of income. Physical capital and land inputs are fixed in the short term, but family and hired labour are variable inputs.

The household-farm sector in our model consists of three groups: commercial, subsistence and net-buyer households. Commercial households are net-surplus producers of agricultural goods. Subsistence households engage in crop production for their own use but typically supplement this with income from non-crop production, wage labour or migration. Net-buyer households have minimal or no involvement in crop production. Household-farm income is the sum of wage income, capital, land and family-labour value-added from household-farm production activities and migrant remittances. Migration from Mexico to the United States and internal migration are a function of the differential between average migrant remittances specific to household groups and the shadow price of family labour in village production activities. The expenditure block includes the consumption demand for village and town products and manufactured goods produced elsewhere in Mexico, leisure, savings, including investments in physical and human capital such as schooling, taxes and some household-to-household transfers.

The general-equilibrium closure equations include village/town market-clearing conditions for factors and goods, a savings/investment balance and a trade-balance equation. For goods and factors for which the village/town economy is a price-taker in regional markets, that is village/town tradables, the market-clearing conditions determine net village/town marketed surplus. For non-tradables, they determine local prices. The savings/investment balance constrains investments in physical and human capital to be self-financed, that is out of local savings. The trade equation constrains the value of village/town exports of goods and factors to equal the value of village/town imports, minus net borrowing from outside the local economy – a village/town analogue to foreign savings in national CGE models. It represents the redundant equation in our village/town CGE system.

Prices of village/town tradables are fixed, determined by regional markets or government policy. Prices of non-tradables are determined by interaction of local supply and demand. Family wages adjust to ensure that family time allocated to production activities, migration and leisure equals total family time allocations.

The price of land is also endogenous, because it is assumed to be fixed and equal to its marginal value product in village/town production activities.

THE VILLAGES AND THE DATA

The research data come from surveys of 50 randomly selected households in each of seven villages and a town located in four Mexican states. The surveys were designed to build databases to support multisectoral and econometric analyses. The surveys gathered the data necessary to calculate production, input use, net incomes from all the main economic activities of each household, expenditures, time use and market and non-market transactions of each household and the relations of the villages and town with regional, national and international markets, and socio-demographic characteristics of household members.

The surveys were designed to mirror the socio-economic structure of Mexico's rural life in several important ways. All villages have fewer than 10 000 inhabitants; the households are involved not only in agricultural and non-agricultural production, but also participate in other income-earning activities in the village and outside it, as in many other parts of rural Mexico. Sample households engaged in agricultural production have characteristics similar to those in other rural locations in Mexico. These are small and medium-sized farms producing maize, beans, cash crops and livestock (Yúnez-Naude, Barceinas and Taylor, 1994).

Most non-agricultural production in a typical Mexican village does not involve manufactures, but includes activities such as processing corn for *tortillas*, handicrafts, services, retail sales, small machinery repairs and making basic construction materials such as bricks. Villages are linked to the outside RNFE primarily through commerce and labour markets. A considerable portion of the demand for inputs and consumer goods in small villages is satisfied by regional production; an important component of villagers' income and sources of investment comes from supplying labour to regional, national or foreign labour markets.

Eight communities were selected from the sample to illustrate farm/non-farm linkages and their implications in different agro-ecological and market environments. The communities include:

- Concordia, an *ejido* or land-reform community in San Pedro de las Colonias, Coahuila;

- El Chante, in Autlan de Navarro, Jalisco;

- Napizaro, Puacurao and Orichu, three villages in the municipality of Erongaricuaro;

- the town of Erongaricuaro, the municipality or county seat, located in the Patzcuaro region of Michoacan;

- Naupan and Reyesoghpan, two indigenous villages located in the mountains of northwest Puebla.

The extreme temperatures and low rainfall of Concordia limit the development of rainfed agriculture and make irrigation indispensable. Until 1992, cotton was the principal commodity produced in La Laguna, where Concordia is located. This changed radically when cotton ceased to be a profitable alternative for La Laguna farmers, and most farmers substituted corn and beans. Basic grain production in La Laguna, including Concordia, is mostly for the market; yields are high and advanced technologies and input packages are used. As demand for labour in cotton production fell, and because of the seasonality of labour demands in maize and bean production, residents of Concordia were forced to seek agricultural and non-agricultural employment in regional towns, in other parts of Mexico, mainly the medium-sized city of Torreón, and in the United States. In short, the economy of Concordia is based on a combination of staple production and local and migration wage work.

El Chante is an agricultural community with a classic cash-crop economy. Rainfed lands are dedicated mainly to maize production for subsistence or for market; irrigated lands are used primarily for sugarcane production. The cane is sold to a mill in the valley, which supplies farmers with inputs and other production resources.

Napizaro is an *ejido* on the shore of Lake Pátzcuaro in one of Mexico's major migrant-sending states. During the last 15 years, its economy has shifted from primarily agricultural staples to livestock. Maize and beans nevertheless continue to have an important role in local production. Most agricultural and livestock production uses family labour, often in combination with purchased chemical inputs and tractor services. Approximately 20 percent of total income consists of income remitted from migrants working in the United States.

Puacuaro and Orichu are very close to Napizaro; households in these two villages also produce staples and livestock. In the town of Erongaricuaro, agriculture remains an important activity along with some light manufacturing and municipal governmental services including the high school. As in Napízaro, remittances from international migrants are a major source of household income. In the SAM multiplier analyses in the third section of the chapter, the three villages are combined into a single village economy; the economies of the villages and the town of Erongaricuaro are then integrated into a model designed to highlight village/town linkages.

Naupan and Reyesoghpan are on steep hillsides of the Sierra; each farmer's agricultural plots are dispersed and very small, 1 ha at most. Of the five villages studied, Naupan and Reyesoghpan have the most indigenous roots. All households in Naupan and Reyesoghpan are engaged in agricultural activities, producing corn, beans and other staples for own consumption, and cash crops. The main cash crop in Naupan is chile; in Reyesoghpan it is coffee. Reyesoghpan farmers are involved in a regional cooperative that provides technological advice and marketing facilities to sell cash crops and to sell goods in a local shop with prices below those of local stores.

The economic structures of the villages: a SAM perspective

The level of economic activity varies in the five villages for which income multiplier experiments are carried out. In terms of gross village output (GVO), El Chante is the largest, producing 1.4 times more than Concordia, the second biggest village. Concordia's GVO is 2.7 times bigger than Naupan; Naupan's GVO is 3.6 times that of Napizaro, whose GVO is 1.6 times that of Reyesoghpan (see Table 1). These rankings do not change when value-added, or gross village product (GVP), is used instead of GVO, but the relative differences in the sizes of the economies change. The gap between El Chante and Concordia widens, for example; the opposite is true for Napizaro and Naupan. This means that El Chante is more internally integrated than Concordia, and that Napizaro is more integrated than Naupan. El Chante's multipliers are hence expected to be greater than those of Concordia, and Napizaro's to be greater than those of Naupan.

Differences in incomes are related to other characteristics of village households, including asset ownership, technology and schooling. Households in El Chante, for example, have an average of more that 7 ha of agricultural land, compared with just over 2 ha for the other non-indigenous villages and fewer than 2 ha for the indigenous villages.[2] Farmers from non-indigenous villages use tractors, whereas farmers in the indigenous villages do not. Average education

---

[2] The differences between the villages increase if we consider land quality is considered. Households in El Chante have an average of 2.4 haectares of irrigated land. Part of the land in the other non-indigenous villages is either irrigated or near a lake, with good water availability, whereas all the land in the indigenous communities is rainfed, on severely eroded steep hillsides with acute erosion.

TABLE 1
Value and distribution of production

1994 Pesos	Concordia	El Chante	Napizaro	Naupan	Reyesoghpan
GVO	11 499 590	16 278 462	1 190 419	4 294 883	735 155
GVP	2 056 131	11 576 696	587 021	1 164 018	520 246
Factor participation in GVP (%)					
Land	10	23	47	8	15
Capital	20	9	30	5	1
Labour	28	14	3	17	13
Family labour	43	53	21	70	72
Sectoral participation in GVO					
Productive activity	23	83	60	27	81
Staples	64	6	17	14	12
Cash crops	0	68	0	59	42
Livestock	12	24	68	7	45
Other agricultural	0	1	11	0	0
Non-agricultural	24	1	4	21	0
Commerce	77	17	40	73	19

Source: Village SAMs and Banco de Mexico, Indices de Precios.

levels range from seven years in Concordia to five years in El Chante and the Municipality of Eronagricuaro and just over 2.7 years in the indigenous villages.

The diversification of income sources typical of rural households is strong in the five villages, including involvement in multiple product and labour markets. Households in all the villages produce staples, cash crops, livestock and other agricultural products such as fruit; in all the villages and the town, commerce is an important activity; in almost all households, some of members are involved in the local, regional, national and international labour markets. By contrast, fewer than 34 percent of the surveyed households are involved in local non-agricultural activities. The composition of non-agricultural production varies between villages. In El Chante, non-agricultural production consists mainly of vehicle and tractor repairs and a brick factory. In Erongarícuaro, the non-agricultural sector is dominated by a small factory producing furniture, a significant source of income for households in the community.

Retail is a major source of household income in all villages; it is as important as agricultural production and more important than non-agricultural production. The exceptions are Concordia and Reyesoghpan, which are located near cities or towns whose shops provide most of the non-staple items required by village households. Other sources of village income include leasing out land, providing capital services such as tractors and household-to-household and government transfers. These are not major sources of income.

The main production activities of the villages are agricultural, but the composition of this agricultural production is different (see Table 1). The most important agricultural activity in Concordia is staple production; livestock raising is the most important agricultural activity in Napizaro. In contrast, cash crops are the most important products of El Chante, Naupan and Reyesoghpan. Another important difference between the villages is local commerce: for Concordia and Naupan it represents more than 72 percent of GVO, for Napizaro 40 percent and for El Chante and Reyesoghpan less than 19 percent.

Family labour is a key production input in all villages, although considerably more important in the indigenous villages in Puebla. The share of family labour in total GVP is about 70 percent in Naupan and Reyesoghpan, whereas it is 54 percent in El Chante, 43 percent in Concordia and 21 percent in Napizaro. Hired labour is a major input in Concordia at 28 percent of value added, less important in Naupan, El Chante and Reyesoghpan, and very low in Napizaro. In Concordia, there is a significant demand for farm labour in the village during the peak season and active participation in regional labour markets. In Naupan and Reyesoghpan there is demand for hired labour during the peak seasons of cash crops. In El Chante, sugar cane production creates a large labour demand, which is satisfied locally and by contracting seasonal migrant workers from the neighboring state of Guerrero. In Napizaro, the most important production activity, livestock raising, uses little labour and virtually no hired labour.

The importance of income from wage labour increases if remittances from family migrants are considered. Remittances by international migrants are the main external source of income for households in the Michoacán microregion. Most remittance income in Concordia and all in the low-income Indian villages comes from family migrants working in other parts of Mexico.

Income sources and their distribution

Sources of household income vary by village (see Table 2). Income from local activities is the most important source in El Chante, Napizaro and Reyesoghpan, where agriculture is the main income source, accounting for about 70 percent household income. Sources of income in Naupan are evenly distributed between local and regional, whereas Concordia's households rely much more than any other village on regional sources of wage income.

The distribution of income between the grouping of households carried out to build the five village SAMs is in Table 3. Concordia and El Chante consist of

TABLE 2
Net income (percent)

	Concordia	El Chante	Napizaro	Naupan	Reyesoghpan
Agriculture	16	68	70	29	74
Crops	83	81	17	90	60
Livestock	17	18	68	10	40
Other agricultural	0	1	15	0	0
Non-agricultural	11	11	7	27	2
Productive activities	21	3	22	23	11
Commerce	79	97	78	77	89
Remittances					
Rest of region	0	0	0	44	8
Rest of Mexico	71	16	4	0	17
International	2	5	18	0	0
TOTAL	100	100	100	100	100

Source: Village SAMs.

TABLE 3
Income distribution by household type

Concordia	%	El Chante	%	Napizaro	%	Naupan	%	Reyesoghpan	%
Agricultural	61	Staple	26	Subsist.	5	Subsist.	34	Subsistence	29
Non-agric.	39	Cash crops	31	Medium	47	Medium	37	Medium	35
		Mixed	21	Large	48	Large	28	Large	36
		Non-agricultural	22						
Total	100		100		100		100		100

Source: Village SAMs.

non-agricultural households; the remaining three villages consist of farming households. The main income sources in non-agricultural households in Concordias and El Chante are wage labour followed by remittances from the rest of Mexico and from the United States. Agriculture is the main income source for subsistence households in Napizaro, although they own little land. The main income source for medium and large farm-households in Napizaro is also farming, but these households own much more land and are engaged in remunerative livestock production. An important portion of the income of these households comes from remittances of family members who are international migrants. The income sources of Naupan and Reyesoghpan are similar across the three household groups. Subsistence households rely more than medium and large farm households on remittances from internal migrants.

Linkages with outside markets

The openness of rural Mexican communities to the outside world is striking. All of them export agricultural and livestock to domestic markets outside the village, receive income in the form of worker remittances from the rest of Mexico or abroad, and import consumption goods and inputs from nearby towns. These market linkages transmit the impacts of changes in outside economies to the communities and vice versa. Village or town imports represent leakages that dampen local income multipliers, or more accurately transfer multipliers resulting from local income changes to other parts of the Mexican economy.

Production

An important portion of village crop and livestock production is sold outside the villages in regional markets. This is true to a lesser extent of non-agricultural production. There are sharp differences between communities, however (see Table 4). Virtually all of El Chante's sugar cane production is sold to a nearby mill; most of its staple production is sold in the region. In Concordia, 93 percent of staple production and 74 percent of non-agricultural output is marketed outside the village. Napizaro is a livestock and migrant exporting economy: 65 percent of livestock output is sold outside the village. Most of the rest is sold locally as investment goods. A large portion of Naupan's cash-crop output and livestock is sold in regional markets; the same is true of Reyesoghpan's cash-crop output.

The destinations of staple production is differ among the communities. Most of the output from Concordia and El Chante is sold in the regional market; most of the staple production in Naupan and Reyesoghpan is for own consumption. The differences between Concordia and El Chante and the other villages in terms of output destination also holds for non-agricultural production.

Inputs purchased for agricultural production activities outside the village represent from 12 percent to 38 percent of the gross value of staple production; the non-agricultural sector typically requires over 50 percent of the non-agricultural sector's GVO. Linkages between the villages and outside product markets also occur through purchases of inputs from stores in the villages, the commerce sector in the SAM, because most commercial-sector goods originate outside the village. Staple producers in Concordia, for example, spend 23 percent of their GVO on inputs from the village commerce sector in addition to 18 percent of their CVO on inputs purchased outside the village. Reyesoghpan's productive activities rely much less on local commerce, because the high proportion of local inputs in the commercial sector results from local intermediaries buying cash crops from farmers to sell outside the village. The single largest production

TABLE 4
Destination of village production (percent)

	Concordia	El Chante	Napizaro	Naupan	Reyesoghpan
Own consumption					
Staple	7	8	40	75	68
Cash crops	0	0	0	1	5
Livestock	97	56	35	26	59
Other agricultural	0	65	74	0	0
Non-agricultural	0	0	52	44	69
Commerce	0	0	0	0	0
Local sales					
Staple	0	18	51	25	32
Cash crops	0	1	0	24	30
Livestock	0	0	0	4	23
Other agricultural	0	35	0	0	0
Non-agricultural	26	0	0	0	0
Commerce	98	96	100	50	23
Outside sales	**Rest of Mexico**	**Region**	**Rest of Mexico**	**Region**	**Region**
Staple	0	74	9	0	0
Cash crops	93	99	0	75	66
Livestock	3	29	65	70	19
Other agricultural	0	0	26	0	0
Non-agricultural	74	100	48	56	31
Commerce	2	4	0	50	77

Source: Village SAMs.

sector across all communities is commerce, which serves as the conduit through which many goods enter the village or town from outside markets (Table 5).

The total value of village imports – goods purchased from domestic markets outside the village or, in rare cases, from abroad – exceeds gross village product in Concordia, El Chante and Naupan. The resulting village trade deficits are financed by income from wage work outside the village through migrant remittances, in other words through labour exports, and by government transfers.

Leakages may take the form of capital outflow from villages, for example to banks in nearby cities. In El Chante, investment leakages from the village are large in relative and absolute terms. Investment outflows make up 90 percent of

TABLE 5

Origin of inputs as a percentage of GVO and weight of inputs in village GVP

	Concordia	El Chante	Napizaro	Naupan	Reyesoghpan
Imported inputs	66	18	43	62	8
Rest of region	N/a	16	na	60	8
Staple	N/a	23	N/a	0	20
Cash crops	N/a	15	N/a	0	3
Livestock	N/a	6	N/a	38	2
Other agricultural	N/a	34	N/a	0	0
Non-agricultural	N/a	0	N/a	0	58
Commerce	N/a	24	N/a	0	23
Rest of Mexico	65	2	43	2	0
Staple	17	0	12	0	0
Cash crops	0	0	0	1	0
Livestock	5	0	13	0	0
Other agricultural	0	0	0	0	0
Non-agricultural	0	0	0	21	0
Commerce	81	13	91	80	0
Local Inputs	16	9	8	11	16
Staple	23	16	13	23	2
Cash crops	0	0	0	20	0
Livestock	27	38	12	15	8
Other agricultural	0	0	0	0	0
Non-agricultural	71	32	55	26	0
Commerce	10	4	0	5	69
Ratio of imports to GVP	3.71	1.25	0.90	2.74	0.81

Source: Village SAMs.

total investment – more than 5.5 million pesos – representing almost 40 percent of leakages from the village. At the other extreme are Concordia, Naupan and Reyesogpan: capital leakages from these villages represent only about 0.5 percent of total investment and a much smaller fraction of leakages.

The communities have strong ties with outside labour markets. For Concordia, the regional labour market in the Laguna region and the nearby city of Torreon accounts for more than 71 percent of village income. Household income in Napizaro includes remittances from migrants elsewhere in Mexico – 4 percent – or in the United States – 19 percent. The weight of international remittances in

Concordia's and El Chante's household income is low; there are no international migrants in the Sierra villages. In Naupan, however, remittances from migrants in the rest of the region represent 44 percent of total income; in Reyesoghpan, remittances from the rest of Mexico account for 17 percent of total income.

The village/town microregion: a SAM perspective

As in the other communities, the economies of Napizaro, Puacuaro, Orichu and Erongaricuaro are diversified. The income sources of their households include involvement in multiple product and labour markets. Households in the three villages and in the town produce staples, cash crops, livestock and other agricultural products; commerce, wage labour and remittances are important in all of them. The composition of their economic structures varies, however. Commerce and services are important in the three villages, representing 34 percent of the villages' GVO; they are even more important in the town, where they account for 50 percent. This difference results from the importance of town stores that supply goods to villagers and villages, and from government services located there. The composition of the productive activities varies among the three villages and the town: livestock is the major activity in the villages, while non-agricultural production is the main productive activity in Erongaricuaro.

The three Michoacan villages are linked to the town of Erongaricuaro through commodity markets. The town is not a source of intermediate inputs for village production activities; these activities obtain their inputs either from within the village, for example local seed and feed for livestock, or else from distributors outside the village/town microeconomy, for example fertilizer. The one exception is services, consisting principally of the village stores that stock their shelves with consumption goods from Kleenex to Wheaties, some of which are supplied by distributors operating out of the town. Households also purchase consumption goods directly from the town, bypassing village stores. Of the total value of goods and services purchased by stores in the villages, 35 percent comes from Erongaricuaro. Of total village merchandise sales, 34 percent are to Erongaricuaro. Through these market linkages, changes in village incomes such as multiplier effects of migrant remittances are transmitted to the town, setting in motion an income multiplier.

Changes in town incomes could in turn stimulate incomes in the villages by increasing the town's demand for village-supplied goods. Backward linkages from town to villages are small, however, especially relative to the size of the town economy. The only such linkages evident in the SAM are for commodity purchases for commercial agriculture, and for the town's nonagricultural sector, primarily wooden furniture produced in Napizaro for a furniture-finishing factory in the town.

Linkages between the town and regional markets outside the village/town microeconomy are crucial in determining the size of town income multipliers resulting from exogenous changes in town or village incomes. The town sub-SAM reveals a high degree of openness to outside markets, particularly for consumption goods entering the town through the service sector – town stores – or through household purchases outside the microregion. The extraregional savings leakage equals about 25 percent of total savings in the villages and 50 percent of total savings in the town; there is no bank in either the villages or the town. Capital outflow from the microregion is small compared with the income leakages created by trade, representing only about 10 percent of total leakages.

FINDINGS FROM VILLAGE AND VILLAGE/TOWN SAM MULTIPLIER MODELS

The village and town SAMs provided the basic data input for the six multiplier models. In the case of the Erongaricuaro region, two SAMs, one for the town of Erongaricuaro and one for the three villages, were stacked into a single SAM matrix with four rest-of-the-world accounts: a local endogenous trade account within the microregion, two other rest-of-Mexico accounts, one for non-competitive intermediate inputs and the other for commodities, and the rest of the world outside Mexico, essentially the United States. Lewis and Thorbecke (1992) combined village and town production activities into a single set of accounts in their SAM multiplier study of a Kenyan village/town economy. Here they are kept separate in order to highlight differential impacts of income changes on production in the two economies. The village and town multiplier models were all programmed using the General Algebraic Modelling System (GAMS) software.

Three experiments were carried out on each of the villages and on the Erongaricuaro region. They were designed to illustrate household income linkages and the implications of agricultural supply constraints for rural income multipliers. The first set of experiments explores the impacts of a US$100 change in exogenous household incomes upon production, factor and household incomes, and trade linkages. This is the least constrained type of SAM multiplier, because it assumes that the output of goods is perfectly elastic; that is, the model is constrained only on the demand side.

Where input supplies are tight or there are short-term capacity constraints, the assumption of perfectly elastic supply response may not be tenable; the model may be constrained on the supply rather than the demand side. In a second set of experiments, agricultural production was constrained to be perfectly inelastic in each of the villages and the village-town economy, as in semi-input-output and economic-base models (see Haggblade, Hammer and Hazell, 1994), using the procedure suggested by Lewis and Thorbecke (1992). We then re-estimated the household income multipliers. In both cases the experiment corresponds to the PROCAMPO programme, in which farmers receive direct income payments per hectare cultivated with basic grains. The first two experiments thus examine the likely impacts of US$100 of PROCAMPO payments. The third set of experiments explores the local economy-wide effects of loosening the agricultural supply constraint by US$100. These experiments might be viewed as simulating the multiplier effects of supply-augmenting technological change in agriculture. The results of all three experiments for the five villages are presented first, followed by the findings for the Erongaricuaro village-town microregion.

Village income multipliers

SAM multipliers without supply constraints

Multiplier effects including farm/non-farm linkages from changes in rural incomes were estimated by increasing total exogenous household income by US$100 and distributing this income across household groups in proportion to their initial shares in total household income in the village. The resulting village SAM multipliers unfold as follows. In the first instance, incomes of the directly affected households increase by the amount of the change in exogenous income. This increases household demands for farm and non-farm goods by the amount of the exogenous income gain times the household-specific marginal expenditure propensities. Because both farm and non-farm production sectors are unconstrained in this model, that is their supplies are perfectly elastic, output increases to meet these increased household consumption and investment demands. This sets in motion a round of production linkages as sectors purchase inputs from one another to support their production expansion. As production increases, new value-added is generated; this value-added is channelled back into households as income, which in turn unleashes a new round of household expenditure increases. Along the way, demands for consumption goods, investment goods and production inputs from outside the village create income leakages, dampening the income multipliers at each round of the process. Although these village imports are a leakage from the point of view of the village, they are mostly non-farm goods and thus represent an important growth linkage between villages and regional or national economies. The SAM multipliers reported in Table 6 represent the end result of multiple rounds of increases in

TABLE 6

Village SAM multipliers resulting from a US$100 increase in rural household incomes[a] (unconstrained model)[b]

	Concordia	El Chante	Napizaro	Naupan	Reyesoghpan
Production					
Staples	2.0	1.5	15.7	8.9	11.5
Cash crops	–	1.7	–	3.6	5.1
Livestock	4.4	15.9	37.2	1.5	35.3
Non-agricultural	2.3	–	12.0	6.9	0.3
Commerce	111.5	17.8	101.0	71.8	5.4
Factors					
Family labour	5.9	17.8	16.8	17.7	30.2
Hired labour	5.2	1.4	2.4	2.9	4.8
Capital	1.7	1.5	20.0	1.4	0.4
Land	0.2	0.9	23.5	2.1	6.0
Household real income	**113.3**	**120.7**	**160.6**	**124.3**	**146.3**

	Agricultural: 66.5	Staple: 32.3	Subsist.: 8.3	Low: 38.5	Low: 40.4
		Cash crop: 37.2	Med.: 76.0	Med.: 47.5	Med.: 49.7
	Non-ag: 46.8	Mixed: 25.3	Large: 76.3	High: 38.3	High: 56.2
		Non-ag: 25.9			

	Concordia	El Chante	Napizaro	Naupan	Reyesoghpan
Savings					
Financial capital	16.5	51.1	19.2	8.7	12.6
Human capital	4.5	4.1	3.6	2.8	0.3
External linkages					
Rest of region		84.8		12.4	65.9
Rest of Mexico	91.2	11.5	102.6	84.7	22.0
Rest of world	1.6	0.0		–	–

[a] Distributed across household in proportion to initial income shares.
[b] All supplies perfectly elastic.

household incomes and production that result from the initial increase in exogenous household income.

The impacts of the US$100 exogenous income change vary strikingly across all villages and across production activities within the villages. Total household income increases by US$113 in Concordia, US$121 in El Chante, US$161 in Napizaro, US$124 in Naupan and US$146 in Reyesoghpan. Much of the impressive increase in income in Napizaro comes from livestock, on which households spend a large marginal share of consumption and investment expenditures. Because livestock production uses little labour, the gains from output expansion in this activity accrue mostly to capital and land. As for Napizaro, most of the increase in Reyesoghpan GVP comes from livestock. Livestock production in the Sierra consists of small animals, however, using family labour. For this reason, the gains from output expansion in Reyesoghpan accrue mostly to family labour.

Apart from livestock, farm/non-farm linkages within the three villages are relatively modest: non-agricultural production rises by just over US$2 in Concordia, by US$12 in Napizaro, by US$7 in Naupan and by US$0.3 in Reyesoghpan; there is no non-farm sector in El Chante. Farm/non-farm linkages outside the villages are striking, however. In all four cases, total demand for goods produced outside the village, mostly manufactured, rises by nearly as much as the amount of the initial exogenous income injection; in one case it is more than this. In Concordia, Napizaro and Naupan most of this external linkage is created by consumer demand for local commercial-sector goods, whereas in El Chante and Reyesoghpan most of it is with the regional products markets. In El Chante, where saving and investment propensities are very high, much of the external linkage is a result of investment demand for outside goods.

The distribution of income gains across household groups is unequal in Napizaro, where most of the indirect income gains go to owners of capital. Income gains are more evenly distributed in Concordia and especially Naupan, Reyesoghpan and El Chante, where most value-added accrues to family labour. In the present experiment an initially unequal income distribution, as in Napizaro, tends to promote an unequal distributional outcome, because the US$100 exogenous income gain is allocated across households according to their initial shares in total income. In all cases, the higher second-round effects of the income change contribute significantly to household incomes. In addition to the initial income injection, for example, the increase in village production stimulated by higher rural household expenditures adds another 62 percent to household income in Napizaro. In general, the households that are most engaged in the production of goods and services for village markets benefit most from the higher second-round effects of the income change.

Village SAM multipliers in the presence of agricultural supply constraints

The size of SAM multipliers resulting from an increase in rural household income depends critically on the supply response of village or town production sectors.

In the extreme case of perfectly inelastic supply response in all production sectors, there can be no income multiplier. In real life, supply elasticities vary across sectors. In agriculture, supply is generally inelastic in the short term. Technological and environmental constraints may result in sharply decreasing marginal returns to factor inputs. If new technologies are not available or accessible to small farmers, the supply response may be inelastic in the long term as well. A number of factors affect the accessibility of new technologies and farmers' willingness and ability to adopt them, including information, risk and the absence of formal or informal insurance markets, lack of access to credit, environmental constraints and various other types of input and output market failures (Feder, Just and Zilberman, 1985; Bellon and Taylor, 1993; Smale, Just and Leathers, 1994; Meng, 1997).

In the second set of experiments, the supply response is constrained in all agricultural sectors to be perfectly inelastic. The exogenous increases in household incomes in this experiment are identical to those in the first experiment. The resulting increase in household demand for village-produced goods, however, now stimulates an increase in output only in the unconstrained non-agricultural sectors. In the agricultural sectors – crop and livestock production – a rise in demand by households or firms can be satisfied only through a decrease in village exports, or marketed surplus. An inelastic supply of agricultural goods constrains the expansion of production sectors that demand inputs from agriculture.

The findings from the constrained-SAM multiplier experiments appear in Table 7. The increases in village income resulting from exogenous rural household income changes are much smaller than in the unconstrained experiments. On the production side, output changes in all agricultural sectors are assumed to be zero. The effect of inelastic agricultural supply on the non-agricultural sectors varies, depending critically upon reliance by the production sectors on the constrained agricultural sectors for inputs. The extreme case is Reyesoghpan, whose economy is the most dependent on agricultural production.

Although increases in household incomes are less than in the previous experiment, the change varies strikingly across villages and household groups. All household groups are hardest hit by agricultural supply constraints in Napizaro and Reyesoghpan, where agricultural production accounts for an important share of household income. In all villages, the household groups most closely linked to the constrained agricultural sectors suffer the most from the supply constraints.

The smaller total income multipliers result in weaker demand linkages between the villages and outside markets. The adverse effects of agricultural

TABLE 7

Village SAM multipliers resulting from a US$100 increase in incomes[a] (constrained model)[b]

	Concordia	El Chante	Napizaro	Naupan	Reyesoghpan
Production					
Staples	0.0	0.0	0.0	0.0	0.0
Cash crops	–	0.0	–	0.0	0.0
Livestock	0.0	0.0	0.0	0.0	0.0
Non-agricultural	2.2	–	8.6	6.4	0.2
Commerce	106.2	11.3	71.6	64.1	3.9
Factors					
Family labour	2.4	6.4	7.1	11.2	0.4
Hired labour	4.8	0.1	1.2	0.6	0.0
Capital	1.3	0.2	7.5	0.9	0.0
Land	0.0	0.0	0.9	0.0	0.0
Household real income	**108.8**	**106.9**	**115.9**	**115.7**	**104.5**
	Agric: 63.6	Staple: 28.1	Subsist.: 6.0	Low: 36.5	Low: 29.4
		Cash crop: 33.5	Med.: 54.7	Med. 44.3	Medium: 35.3
	Non-ag: 45.2	Mixed: 22.2	Large: 55.2	High: 34.9	High: 39.8
		Non-ag: 23.1			
Savings					
Financial capital	15.8	45.0	13.9	8.1	9.0
Human capital	4.3	3.6	2.6	2.6	0.2
External linkages					
Rest of region		71.6		10.2	44.2
Rest of Mexico	86.3	9.3	68.0	75.9	15.7
Rest of world	1.5	0.0	–	–	

[a] Distributed across household in proportion to initial income shares.
[b] Agricultural production sector supplies perfectly inelastic.

supply constraints are larger for village trade than for village incomes. In short, the capacity of changes in farm incomes to stimulate demand for non-farm goods turns critically on the agricultural supply response.

Village multipliers from loosening agricultural supply constraints

Loosening agricultural supply constraints can have a multiplicative effect on village incomes and stimulate farm/non-farm linkages. The third set of experiments was designed to estimate agricultural supply multipliers in the five villages. To do this, the constrained SAM of the previous experiment was used as the base model; agricultural output was then exogenously increased by US$100, with this increase distributed across agricultural production sectors in proportion to their initial shares in total agricultural output. This experiment might represent a loosening of supply constraints through technological change, credit market development or public investments in marketing, extension or transportation infrastructures. Alternatively, by changing the sign of the supply shift from positive to negative, it might represent the negative multiplier effect of environmental degradation diminishing production.

The increase in agricultural supply has several impacts on a village economy. It increases intermediate input demands by the constrained agricultural sectors, stimulating production and imports of these inputs and setting in motion a round of production linkages in the village. It also increases payments of value-added income to village households, which results in a rise in household expenditures on goods and services supplied inside and outside the village. The household demands stimulate a new round of production increases in the village, increased demands for intermediate inputs and value-added payments to households. The SAM multipliers resulting from the increase in agricultural-sector supply are presented in Table 8. The exogenous increases in agricultural-sector supply appear in the first three rows of the table.

This experiment produces by far the largest increases in village value-added, reflecting the pivotal role of agricultural supply constraints in shaping rural income linkages. In this experiment, all village income effects are indirect, through production. The increased agricultural production generates a large increase in labour value-added in the four most labour-intensive villages; in Napizaro, where most of the output gain is in the capital-intensive livestock activity, physical capital and land value-added represent most of the increase in value-added.

It is not surprising that agricultural households benefit disproportionately from the supply shift. The fact that non-agricultural households also benefit, although relatively modestly (see Concordia and El Chante columns), reflects the existence of some farm/non-farm commodity and factor-market linkages within these villages. As in the previous experiments, the distributional effect on household incomes is much more equitable in Concordia, El Chante, Naupan and Reyesoghpan than in Napizaro, again reflecting the capital intensity of Napizaro livestock production and an unequal distribution of capital in that village.

TABLE 8

Village SAM multipliers resulting from a US$100 increase in agricultural supply[a] (constrained model)[b]

	Concordia	El Chante	Napizaro	Naupan	Reyesoghpan
Production					
Staples	84.3	6.3	20.2	17.5	12.2
Cash crops	–	69.6	–	74.0	42.7
Livestock	15.7	24.2	79.8	8.5	45.0
Non-agricultural	1.3	–	6.3	3.4	0.2
Commerce	69.3	15.5	54.2	58.6	3.2
Factors					
Family labour	31.7	43.5	16.6	51.0	61.5
Hired labour	9.5	12.1	1.9	18.8	10.9
Capital	15.2	8.0	23.9	2.7	0.5
Land	9.9	19.6	44.3	9.8	12.8
Household real income	**66.4**	**70.7**	**83.7**	**61.5**	**87.7**

	Agric:	44.0	Staple:	15.6	Subsist.: 4.3	Low:	12.8	Low:	23.5
			Cash crop:	26.1	Med.: 40.0	Med.:	23.5	Medium:	30.2
	Non-ag:	22.4	Mixed:	16.5	Large: 39.4	High:	25.2	High:	34.0
			Non-ag.:	12.5					

	Concordia	El Chante	Napizaro	Naupan	Reyesoghpan
Savings					
Financial capital	9.8	30.0	9.9	4.8	7.6
Human capital	2.7	2.4	1.9	1.2	0.2
External linkages					
Rest of the region		74.7		13.5	43.1
Rest of Mexico	71.9	9.0	64.3	64.9	13.2
Rest of world	1.0	0.0	0.0	0.0	0.0

[a] Distributed across agricultural production sectors in proportion to total initial output levels.
[b] Agricultural production sector supplies perfectly inelastic.

The increase in agricultural supply has modest positive effects on non-farm production in the five villages, but strong effects on village demands for manufactures. Again, the increase in local commerce is much lower in the villages that are well connected with the regional product markets, El Chante and Reyesoghpan.

Income multipliers in the village/town microregion

The findings reported above reveal modest farm/non-farm income linkages within Mexican villages. The Erongaricuaro microregion survey affords the opportunity to estimate farm/non-farm linkages between village economies and the town economy to which they are most directly linked. For each variant of the SAM multiplier model presented for the villages, two sets of experiments were performed for the village/town microregion. The first involves an exogenous US$100 increase in village household incomes for the unconstrained and constrained cases, or an exogenous US$100 increase in constrained-sector supply for the constrained case. These experiments were repeated for the town economy. The multiplier effects of these two experiments on village and town incomes are summarized in Table 9, Table 10 and Table 11. Columns A and B in these tables present multiplier impacts on the village and town economies that result from a US$100 exogenous increase in village-household incomes (Table 9) or in constrained-sector supply (Tables 10 and 11). Columns C and D present impacts on village and town economies associated with the same exogenous increases in town-household incomes or supply.

Unconstrained village/town SAM multiplier

The multiplier effects of the exogenous income increase are substantially larger in the villages than in the town; for example, total income increases by US$231 as a result of the US$100 exogenous increase in village household incomes (column A). By contrast, town income rises by only US$133 as a result of the US$100 exogenous increase in town household incomes (column D). The much smaller value-added multiplier for the town can be explained by the town's high degree of openness to markets outside the village/town region. These outside market linkages transfer the multiplier out of the village/town microregion. The villages, on the other hand, are less integrated with outside markets; a large proportion of changes in demand is satisfied through changes in village production. In the villages (column A), the exogenous income gain is associated with production increases that dwarf the production increases in the town associated with a US$100 exogenous income change there (column D). In both town and village economies the largest increases are for commerce, which mainly represents a leakage from the local economy.

The cross effects of the village income increase on the town economy appear in column B; the town income increase on the village economy is shown in column C. In all cases, cross effects are small relative to the own effects. Because the town's commercial sector plays an important role in satisfying village

TABLE 9
Village/town SAM multipliers resulting from a US$100 increase in rural household incomes[a] (unconstrained model)[b]

Increase in household income	A Villages	B Town	C Villages	D Town
Production				
Basic grains	23.2	0.5	0.0	4.6
Other grains	14.1	1.2	0.1	7.8
Livestock	52	1.7	0.1	8.7
Ren res.	13.3	0.1	0.0	2.5
Non-agricultural	17.5	0.1	0.2	2.8
Commerce	73.6	38.4	0.2	17.0
Factors				
Family labour	71.9	5.0	0.3	15.8
Hired labour	15.8	2.0	0.1	2.2
Physical capital	9.1	0.1	0.0	0.5
Animal capital	1.0	0.1	0.0	0.5
Land	17.5	0.3	0.0	2.8
Household real income	**231.1**	**8.0**	**0.4**	**133.2**
Commercial	102.4	3.6	0.2	44.7
Subsistence	71.5	1.1	0.1	26.0
Net buyers	57.2	3.3	0.1	62.5
Savings				
Financial/physical capital	18.6	0.9	0.0	15.5
Human capital	6.3	0.7	0.0	14.4
External Linkages				
Rest of Mexico	5.0	0.1	0.0	0.9
Rest of Mexico commodities	17.5	5.5	0.1	72.2
Rest of world	40.4	31.6	0.1	26.7

[a] Distributed across household in proportion to initial income shares.
[b] All supplies perfectly elastic.

consumption demands, it is the town's main beneficiary from increases in village incomes. The cross effect of village income increases on town commerce of US$38 actually exceeds the US$17 effect of a US$100 increase in the town's income on its own commerce.

TABLE 10

Village/town SAM multipliers resulting from a US$100 increase in incomes[a] (constrained model)[b]

Increase in household income	A Villages	B Town	C Villages	D Town
Production				
Basic grains	0.0	0.4	0.0	0.0
Other grains	0.0	0.6	0.0	0.0
Livestock	0.0	1.1	0.0	0.0
Ren res.	8.2	0.1	0.0	2.2
Non-agricultural	10.7	0.1	0.1	2.5
Commerce	45.2	23.3	0.0	13.8
Factors				
Family labour	22.4	3.0	0.1	4.3
Hired labour	7.3	1.2	0.0	1.5
Physical capital	3.0	0.1	0.0	0.1
Animal capital	0.0	0.0	0.0	0..0
Land	0.0	0.2	0.0	0.0
Household real income	**142.6**	**4.9**	**0.1**	**116.0**
Commercial	59.7	2.2	0.1	35.6
Subsistence	44.4	0.7	0.0	22.8
Net buyers	38.5	2.0	0.0	57.6
Savings				
Financial/Physical capital	11.6	0.53	0.0	13.9
Human capital	3.8	0.4	0.0	12.8
External linkages				
Rest of Mexico	0.0	0.1	0.0	0.0
Rest of Mexico commodities	4.1	3.3	0.0	61.7
Rest of world	25.4	19.2	0.0	22.4

[a] Distributed across household in proportion to initial income shares.
[b] Agricultural production sectors supplies perfectly inelastic.

In the village/town SAM, there are three household groups. Commercial households are household-farms or household-firms engaged in commercial production of some sort. Subsistence households are households that produce for their own subsistence only. Net-buyer households include households that satisfy little or none of their consumption demand from own production. Commercial households enjoy the largest income gains; they benefit directly from the exogenous income gain and indirectly through increased demand for the goods they produce. Taking into account the multiplier effects of the exogenous income change, the income of all household groups more than doubles. The large divergence between total and direct effects on household incomes reflects the involvement of households in production activities stimulated by income linkages in the village/town microregion.

In the town, net-buyer households have the highest incomes to begin with and thus initially benefit most from the income transfer. As in the villages, the indirect effects of the income change are largest for commercial households, though overall indirect effects are more muted. They are smallest for subsistence households.

Although the income change stimulates a village demand for town commerce, external linkages with the rest of Mexico are relatively small. The US$100 exogenous income increase in the villages increases village demand for goods from the rest of Mexico by just over US$22. By contrast, the US$100 increase in town incomes increases town demand for goods from the rest of Mexico by US$73. Village households spend a significant part of their income gains on goods brought home by migrants from the United States; the demand linkage with the United States is smaller for the town.

Village/town constrained multipliers

Multiplier effects of the exogenous income change plummet when the output of agricultural and livestock activities is constrained. The effect of the US$100 exogenous income change in the village on village income falls from US$231 in the unconstrained case to US$143; the effect in the town on town household incomes falls from US$133 to US$116. The cross-multipliers decrease as well, although in the case of the village-to-town effect the decrease is much smaller in relative terms than the decrease in village and town own multipliers.

By assumption, there is no change in production in the output-constrained agricultural and livestock sectors in the village. To compensate for these constraints, the village economy shifts its demands away from village supply-constrained goods in favour of unconstrained-sector goods, including those supplied by the town. As a result, the output of unconstrained sectors decreases by less in relative terms than the decrease in total village value-added; so does the multiplier effect on gross town product. In short, supply constraints in the village substantially dampen the village income multiplier. The negative effects

TABLE 11

Village/town SAM multipliers resulting from a US$100 increase in agricultural supply[a] (constrained model)[b]

Increase in household income	A Villages	B Town	C Villages	D Town
Production				
Basic grains	25.4	0.3	0.1	17.8
Other grains	14.8	0.5	0.2	26.1
Livestock	60.0	0.8	0.4	56.1
Ren res.	5.7	0.1	0.1	1.3
Non-agricultural	7.8	0.1	0.7	1.6
Commerce	32.5	17.3	0.4	15.0
Factors				
Family labour	56.8	2.2	0.9	61.6
Hired labour	9.9	0.9	0.2	3.3
Physical capital	6.7	0.1	0.1	2.1
Animal capital	1.0	0.0	0.0	1.6
Land	19.8	0.2	0.1	12.2
Household real incomes	101.1	3.6	1.3	86.9
Commercial	48.8	1.6	0.6	46.3
Subsistence	30.9	0.5	0.4	15.0
Net buyers	21.4	1.5	0.3	25.6
Savings				
Financial/physical capital	8.0	0.4	0.1	8.4
Human capital	2.9	0.3	0.0	8.0
External linkages				
Rest of Mexico	5.3	0.0	0.0	3.1
Rest of Mexico commodities	14.5	2.5	0.3	52.1
Rest of world	17.1	14.2	0.2	20.9

[a] Distributed across agricultural production sectors in proportion to total initial output levels.
[b] Agricultural production sector supplies perfectly inelastic.

of village production constraints on the town economy are smaller, however, as village production activities and households attempt to compensate for the supply constraints by shifting their intermediate and consumption demands in favour of town-produced goods.

The decrease in village income and production multipliers is reflected in household incomes, which increase by about 40 percent less than in the unconstrained case. Agricultural supply constraints substantially reduce the indirect effects of the income change, especially for commercial households, which are heavily involved in the constrained-sectors production. As in the first experiment, commercial households enjoy the largest gains from the exogenous income change in the village. The ratio of indirect to direct income changes drops from 149 percent to 46 percent for commercial households, however; in the supply-constrained case it is similar to the 44 percent of village net-buyer households. In the town, the ratio of indirect to direct income effects for commercial households falls from 40 percent to 12 percent.

Village multipliers from loosening agricultural supply constraints

Loosening agricultural supply constraints unleashes powerful income multipliers in the village/town economy, as in the village economies examined earlier. In contrast to the effects of the exogenous income increases in the first two experiments, increases in the exogenous supply of agricultural and livestock activities has a similar impact on the village and town economies. A US$100 increase in constrained village agricultural and livestock supply stimulates a US$101 increase in total income in the village. The same exogenous increase in town supply results in an US$87 increase in town income, a much larger indirect effect than in the other two experiments. The constrained-production multiplier is less than 20 percent larger in the villages than in the town; by contrast, the income multipliers reported above were of the order of five times larger in the village. This finding highlights the strong local linkages emanating from agricultural and livestock activities within the village/town economy. By targeting these sectors, the US$100 increase in supply exploits these linkages directly, thus creating larger income multipliers than those generated by the US$100 increase in household incomes, which affect the agricultural and livestock sectors indirectly through household expenditures.

Because households do not benefit directly from the exogenous increase in supplies from agricultural and livestock activities, all the household-income multipliers are indirect in this experiment and the impacts on households' total incomes are smaller than in the other two experiments. The households most involved in agricultural and livestock activities benefit disproportionately from the supply change. This is the only experiment in which town commercial households have an advantage over net-buyer households because of their involvement in agricultural and livestock activities.

VILLAGE/TOWN MICROREGION CGE RESULTS

How sensitive are these findings to the assumptions of perfectly elastic – or perfectly inelastic in the case of the constrained multipliers – supply response, linearity and non-binding resource constraints implicit in SAM multiplier models? A CGE model was used for the village/town economy to explore this question. The model was first estimated on the basis of information in the village/town SAM as the basic data input. The model was then used to simulate the impact of (a) exogenous household income transfers and (b) increases in agricultural supply response in the nonlinear, flex-price environment depicted in the CGE. These experiments were performed separately for each of two market scenarios. Scenario I assumes that there are perfect markets for commodities and factors and that all goods are tradables. Scenario II treats all goods and factors as non-tradables with local endogenous prices. These two scenarios bracket the possible impacts of the experiments on the village/town economy.

Implications of market closure assumptions

Scenario I: The neoclassical village/town with tradables

In a neoclassical world with perfect markets, all goods are tradables; the village/ town microeconomy faces no transaction costs and is a price taker in all commodity and factor markets. This means that supplies of commodities and factors to village/town households and production sectors are perfectly elastic. Unlike the fixed-price SAM multiplier models, however, this perfectly elastic supply comes from markets outside the village/town. Within the village/town economy, household-farms and firms face resource constraints such as fixed capital and technological constraints, resulting in decreasing marginal returns to factor inputs and rising marginal costs of production; that is, their supply curves slope upward. In short, village/town production sectors act like individual agricultural households in a perfect-markets model (see Singh, Squire and Strauss, 1986), taking prices exogenously determined in outside regional or national markets and following first-order conditions for profit and utility maximization.

Because production and consumption decisions in the village/town economy do not influence regional prices, the entire village/town model, like its neoclassical household-farm model counterpart, is separable or recursive. Policy or market shocks influencing the production side of the economy affect consumption through the budget constraints. Exogenous changes in household incomes, however, influence consumption but do not affect village/town production, because under the assumption of exogenous prices the first-order conditions for profit maximization remain unchanged. These income changes do, however, affect net village/town marketed surplus, or trade with markets outside the village-town economy. In a recursive village/town model, exogenous changes in household incomes like those in our first two sets of SAM multiplier experiments do not generate any income multipliers within the village/town economy. The potential multiplier leaks out of the local economy immediately, through regional trade.

Scenario II: The village/town with non-tradables

The creation of multipliers in a CGE model requires price changes that transmit the impacts of exogenous income changes through the production side of the economy; that is, the presence of local non-tradables. In a SAM multiplier model, all demand for locally produced goods is implicitly assumed to be for non-tradables. Prices for these non-tradables are assumed to be fixed by excess capacity in the local economy, which results in perfectly elastic supply of commodities and factors.

Scenario II represents a CGE analogue to the SAM modelling assumption that all goods are non-tradables in the village-town economy. It corresponds to a village/town economy facing high costs of transacting with outside markets for goods and factors demanded in the local economy. It does not rule out trade with these markets; as in the SAM multiplier models, each production sector and some households satisfy a fixed share of their input and consumption demands from outside markets. These import shares represent a leakage from the local village-town economy into the rest of the world.

The fundamental difference between the CGE model in Scenario II and the SAM multiplier models used previously is that in the CGE model, prices of nontradables are endogenous; the village/town economy faces resource constraints that result in upward-sloping supply curves for all production sectors. The village/town model is no longer separable or recursive. Exogenous changes in household incomes, which shift the budget constraints outward, alter household demands for local non-tradables. This would be depicted graphically as an outward shift in local demand curves for normal goods. Given upward-sloping supply curves, the result is an increase in local prices of non-tradable commodities and factors. These price changes link the exogenous changes in household incomes to village/town production sectors, changing the first-order conditions for profit maximization. Local production expands to satisfy increases in household demands, increasing its demand for intermediate and factor inputs and generating new rounds of production and household-income increases. The result is a village/town CGE multiplier, analogous to the SAM multipliers presented earlier.

The findings from the village-town CGE experiments are presented in two parts: first, the household-income transfer experiments; second, the simulated increases in agricultural and livestock productivity.

Household income experiments

Scenario I: The neoclassical village/town with tradables

Table 12 shows the impacts on the village/town economy of a US$100 exogenous increase in household incomes, distributed across households in proportion to their initial income levels. This experiment is identical to the SAM income-transfer experiment shown in Table 11. The first two columns of the Table present impacts on the village first, then the town if the income increase goes to village households. The third and fourth columns report the impacts of an exogenous increase in town-household incomes.

The figures in this table illustrate clearly the assumptions implicit in the perfect neoclassical village/town model. Because all goods are tradables and all prices exogenous, the impacts of the household-income increase are not transmitted to the production side of the economy, and there is no change in village or town real gross domestic products (GDPs). The village income transfer, the first two data columns in the table, directly favours village commercial households, whose initial incomes are higher. These increases in village household income result in higher demands for normal goods. Without an accompanying increase in local production, these increased demands simply reduce the supply of goods available outside the village/town economy. Village marketed surplus falls by US$28. Because part of the increased village demand is satisfied by the town, marketed surplus there falls as well, by US$23. In this scenario of exogenously determined prices, the village/town consumption linkages have no effect on production or household income in the town.

The effects of increased town incomes, the second and third data columns of Table 12, are qualititatively identical to those of increased village incomes; as in the SAM experiments, however, linkages are smaller from town to village than the other way round. Nearly half of the exogenous income change in the town goes to net-buyer households, which are the highest-income household group there. As a result of increased consumption demands by households, town marketed surplus falls by US$22 and village marketed surplus falls by less than US$1.

TABLE 12

General equilibrium effects of a US$100 increase in household incomes[a] (all goods tradable)

Income increase in	Village		Town	
Effects on accounts	Village	Town	Village	Town
Production	0.0	0.0	0.0	0.0
Basic grains	0.0	0.0	0.0	0.0
Other crops	0.0	0.0	0.0	0.0
Livestock	0.0	0.0	0.0	0.0
Renewable resources	0.0	0.0	0.0	0.0
Non-agricultural	0.0	0.0	0.0	0.0
Commerce	0.0	0.0	0.0	0.0
Factors				
Family labour	0.0	0.0	0.0	0.0
Hired labour	0.0	0.0	0.0	0.0
Physical capital	–	–	–	–
Animal capital	–	–	–	–
Land	–	–	–	–
Household real income	100.0	0.0	0.0	100.0
Commercial	40.7	0.0	0.0	32.3
Subsistence	32.6	0.0	0.0	21.7
Net buyers	26.7	0.0	0.0	46.0
Savings	**10.8**		**23.2**	
Financial capital				
Human capital				
Trade				
ROW market surplus	-28.0	-7.4	-0.6	-22.5

[a] Distributed across households in proportion to initial income shares.

Prices are those of commercial and subsistence activities, where applicable.

Scenario II: The village/town with non-tradables

In contrast, exogenous household-income changes generate significant production and income multipliers in the village/town economy when all goods are assumed to be non-tradables. The figures in Table 13 reveal the extent to which price changes transmit transfer-induced changes in household incomes to village/town production sectors. In most cases, the production impacts of the income changes from the CGE model lie in between those for the unconstrained and constrained SAM multiplier models shown in Tables 9 and 10. Among production sectors,

TABLE 13

General equilibrium effects of a US$100 increase in household incomes[a] (all goods non-tradable)

Income increase in	Village		Town	
Effects on accounts	Village	Town	Village	Town
Production	0.0	0.0	0.0	0.0
Basic grains	7.6	1.8	-0.1	3.2
Other crops	3.8	1.5	0.6	0.0
Livestock	26.2	6.6	-1.0	6.0
Renewable resources	9.6	0.5	-0.0	2.2
Non-agricultural	10.1	0.5	0.5	2.8
Commerce	41.6	27.5	-4.6	10.9
Factors				
Family labour	52.7	20.7	-1.0	19.6
Hired labour	11.7	8.0	-0.5	4.5
Household real income	124.0	8.5	-12.0	111.9
Commercial	56.7	-0.1	-6.6	40.3
Subsistence	36.3	2.6	-2.4	21.1
Net buyers	31.0	6.0	-3.1	50.5
Savings	**26.5**		**28.6**	
Financial capital				
Human capital				
Trade				
ROW market surplus	0.0	0.0	0.0	0.0

[a] Distributed across households in proportion to initial income shares.
Prices are those of commercial and subsistence activities, where applicable.

the largest beneficiaries of the village income transfer is commerce. Higher village demand for commercial-sector goods generates a positive income/growth linkage with the town, because village households and production activities spend income in the town, unleashing an endogenous price multiplier there.

Real income in the village/town economy increases substantially as a result of the exogenous increase in village-household incomes; the impact is dampened by higher village-town prices for nontradables, however. In the village, total real income rises by US$124, less than in the constrained and unconstrained SAM models. Village-town income linkages stimulate a US$9 increase in town incomes, more than in either the constrained or unconstrained SAM models. Commercial households benefit most, directly and indirectly, in real terms from the income change. The smallest beneficiaries of the income shock in the village are the net-buyer households. In the town, real incomes increase for subsistence and net-buyer households, but they fall slightly for commercial households.

The openness of the town economy results in a leakage of most of the potential multiplier effects of income changes into the outside world. The US$100 increase in town incomes, the third and fourth data columns of Table 13, raises town real income indirectly by US$12, a respectable increase but small compared with the village income effect of the increase in village incomes, and not much different from the impact of village income increases on town income. The impact on total town income is smaller than in either of the two corresponding SAM multiplier experiments. As in the SAM experiments, the largest beneficiaries of the exogenous income increase are net-buyer households in the town. The largest indirect real benefits, however, accrue to commercial households. Subsistence households lose in real terms as a result of higher prices for consumer non-tradables.

To summarize, the village/town CGE model with non-tradables produces estimated income impacts of exogenous income changes that are in most cases smaller than those predicted by the constrained (low) and unconstrained (high) SAM multiplier models. Nevertheless, in a few cases the non-linearities and relative price changes permitted in the CGE model result in village/town sectoral impacts that are larger than in the fixed-price SAM models; in other cases estimated real impacts are negative in the CGE model, a plausible result that is ruled out in SAM models by construction.

Technology-change experiments

One of the advantages of a CGE approach is that sectoral supply responses are assumed to be neither perfectly elastic nor perfectly inelastic. Because of this, the experiments do not exactly replicate the exogenous increase in agricultural and livestock supply in the constrained SAM model using the CGE approach. Instead, the village/town economy-wide implications of a technological change is explored that initially increases production in the agricultural and livestock sectors by US$100, distributed across these sectors in proportion to their initial production levels. In the first instance, output in the affected sectors increases, shifting the supply curves outward. In the model with non-tradables, this drives down local prices for non-tradables. In both models, payments to factors and households increase, which together with a higher demand for intermediate inputs shifts the local demand curves outward. When goods and factors are non-tradables, upward pressure is put on local prices. In the CGE model with non-tradables, the ultimate effect on local prices is generally indeterminate.

Scenario I: The neoclassical village/town with tradables

The results of the technological change, estimated using the perfect neoclassical variant of the village/town CGE model, are shown in Table 14. The neoclassical model appears to exaggerate the impacts of the technological change on the village/town economy. In the village technological change experiment, village/town real income rises by US$308, more than three times the initial production increase resulting from the technological change. Because livestock output is highest to begin with, this sector receives most of the direct benefit of the technological change. All non-agricultural sectors in the villages and all agricultural and non-agricultural sectors in the town are unaffected by the technological change in this model, because there are no endogenous prices to transfer the impacts to them. Most of the increase in value-added accrues to family labour. Unlike the income transfer experiments, the initial impacts of the technological change experiment are on the production side, and commercial households are most closely linked to these activities. Because there are no price linkages to stimulate town production in this model, town incomes are unaffected by the technological change in village agriculture.

The marketed surplus effects of the technological change are ambiguous. On one hand, production of agricultural and livestock goods increases. This induces increases in household incomes, however, which in turn increases the demand for normal goods in the village/town economy. The net impact on marketed surplus equals the change in village supply minus the change in village/town demand for each good.

Net marketed surplus increases slightly by US$5 for staples, decreases by US$40 for other crops and increases sharply by US$230 for livestock. For all other village and town goods, net marketed surplus decreases, because demand rises while production remains unchanged. The largest decrease in town marketed surplus is predictably in the commercial sector, where the increased village demand results in a US$25 increase in purchases from outside markets. These represent a leakage in the village/town economy but a potential growth linkage elsewhere in Mexico.

Impacts of technological change in the town are similarly exaggerated. Total income in the village/town economy increases by US$550, driven by a huge increase in town livestock output. As in the village technological change

TABLE 14

General equilibrium effects of an increase in productivity[a] (all goods tradable)

Income increase in	Village		Town	
Effects on accounts	Village	Town	Village	Town
Production				
Basic grains	44.3	0.0	0.0	46.1
Other crops	15.0	0.0	0.0	108.0
Livestock	312.6	0.0	0.0	560.6
Ren. res.	0.0	0.0	0.0	0.0
Non-agricultural	0.0	0.0	0.0	0.0
Commerce	0.0	0.0	0.0	0.0
Factors				
Family labour	21.5	0.0	0.0	460.0
Hired labour	12.4	0.0	0.0	7.6
Household real income	307.9	0.0	0.0	550.3
Commercial	156.5	0.0	0.0	325.6
Subsistence	96.6	0.0	0.0	92.3
Net buyers	54.8	0.0	0.0	132.4
Savings	**33.1**		**98.1**	
Financial capital				
Human capital				
Trade				
ROW market surplus	161.3	-26.0	-22.4	307.1

[a] Distributed across activities in proportion to initial production shares.

experiment, commercial households receive the bulk of the income gain; net-buyer households also gain, however, by virtue of their supply of factors to agricultural and livestock production.

Scenario II: The village/town with non-tradables

The results of the technological change experiments are more plausible and closer to those from the SAM experiments where all locally produced goods are treated as non-tradables (Table 15). In both town and village, agricultural and livestock production increases while the impacts on other production sectors tend to be small or negative. They benefit from an income-induced increase in demand for their output and from lower prices of agricultural and livestock inputs. They lose, however, when there is an increase in the prices of the non-tradable factors they employ.

TABLE 15
General equilibrium effects of an increase in productivity[a] (all goods non-tradable)

| Income increase in | Village | | Town | |
Effects on accounts	Village	Town	Village	Town
Production				
Basic grains	22.4	0.1	-0.4	14.1
Other crops	8.6	0.1	1.0	19.3
Livestock	40.5	0.4	1.4	29.0
Ren. res.	0.6	-0.0	0.2	0.4
Non-agricultural	0.1	-0.1	-0.2	-0.4
Commerce	20.7	3.5	14.2	28.8
Factors				
Family labour	-17.2	2.8	6.3	-11.1
Hired labour	0.0	0.0	0.0	0.0
Household income	96.6	-3.6	32.4	146.4
Commercial	25.1	-0.6	19.2	38.9
Subsistence	41.3	-1.0	7.4	47.3
Net buyers	30.2	-2.1	5.8	60.3
Savings	-1.1		3.1	
Financial/physical capital				
Human capital				
Trade				
ROW market surplus	0.0	0.0	0.0	0.0

[a] Distributed across households in proportion to initial income shares.

Much of the impact of technological change is on agricultural and livestock prices; the terms of trade turn sharply against these sectors. This results in large increases in real household incomes for all village and town households. There is a small positive linkage effect from technological change in the town to village production of livestock, other crops and especially services.

CONCLUSIONS

Our SAM and CGE analysis of multipliers resulting from changes in rural incomes and agricultural productivity points to three important conclusions. First, the size of village and village/town income multipliers is potentially large. A US$100 injection of exogenous income into village households stimulates increases of as much as US$63 in gross village product in Napizaro and US$123 in the village/town economy (unconstrained SAM models). The total direct-plus-indirect effects on household incomes reach US$161 in Napizaro and US$231 in the village/town economy. Both the magnitudes of these multipliers and the distribution of income gains across household groups and production sectors are sensitive to village and village/town economic structures. In general, the more open the economy, the smaller the income multipliers. The smallest multipliers estimated using SAM models are in Concordia, which sells a large share of its agricultural output to markets outside the village and satisfies a large share of its input, consumption and investment demands from these outside markets. The largest and most unequally distributed multipliers are in Napizaro, which relies heavily on migrant labour markets for income, but which satisfies an important part of consumption and investment demand locally, particularly from a burgeoning and concentrated livestock-production activity.

Second, farm/non-farm demand linkages are important. The great majority of these linkages, however, are with markets outside rather than within villages. The US$100 increases in exogenous household income stimulated increases in village non-farm production between US$0.3 and US$12, while village demand for manufactures from outside markets increased by US$88 to US$103. In the village/town microregion, a US$100 increase in exogenous income stimulated a US$18 increase in village non-agricultural production, a US$74 increase in commerce demand, almost entirely goods bought outside the village, a US$38 increase in village demand for goods sold in the town and a US$22 increase in village demand for goods from outside the village/town economy. These findings offer compelling evidence in support of what Adelman (1984) calls "agriculture-demand-led-industrialization". In rural Mexico, a large share of rural household demand for purchased inputs and consumption and investment goods is supplied by regional towns, which have proliferated in the last decade and which now account for most of the country's urban growth. As these findings illustrate, most of the farm/non-farm diversification in rural Mexico is between villages, where agriculture is still the economic mainstay, and these growing regional towns and cities. Nevertheless, village household incomes are diversified away from agriculture, largely as a result of families' participation in labour markets outside the village, through wage work or in distant towns or abroad, through migration.

Data from the economic census show that between December 1989 and December 1993, real values of fixed assets grew rapidly in the commercial sector of cities located near the villages and town studied here. Trade linkages transfer most of the benefits of income growth in villages to these regional commercial centres. This does not mean, however, that there is necessarily a parasitic relationship between villages and towns, or between villages and the outside world. Town markets are critical to support village crop and non-crop production activities that create value-added for village households, and regional and extraregional labour markets are a major source of wage and remittance income for villages. Our analysis highlights the complex economic interactions between villages and towns in what is probably a mutually beneficial relationship broadly consistent with comparative advantage.

Increasing the income of village households and loosening agricultural supply constraints is important for the growth of the RNFE in towns and small cities. This is illustrated by our results for the microregion. The town commercial sector plays an important role in satisfying village demands for non-agricultural goods. It is a major beneficiary from rising village incomes and reducing village agricultural-supply constraints. In these experiments, the cross-over effect of village income or agricultural supply on town commerce exceeds the own effect of increased town income or agricultural supply. Positive village/town growth linkages, in turn, increase the town's demand for goods produced in regional cities.

Third, estimated farm/non-farm linkages resulting from rural income changes appear to depend critically on the supply response of agriculture and on model specification, especially the role of prices. In the second set of SAM experiments, it was found that the multiplier effects of changes in rural incomes on production, value-added and village demand for manufactures were considerably smaller when agricultural-supply response was inelastic.

In the arguably more realistic CGE models, supply is assumed to be neither perfectly elastic nor perfectly inelastic: producers face rising marginal costs and thus have upward-sloping supply curves, and prices of local non-tradables change in response to exogenous income and policy shocks. When some goods or factors in the local economy are non-tradables, endogenous price changes transmit the impacts of income changes from households to the production side of the local economy, creating potentially large local production and income multipliers. The village/town CGE model with all goods non-tradable produced estimated impacts of production changes that tended to be in between those predicted by the constrained and unconstrained SAM models, with estimates of total income changes that were large but smaller than those obtained using the unconstrained SAM. That is, estimated total income linkages resulting from exogenous changes in rural incomes tended to be unaffected by the choice of model.

Estimates of the distributional consequences of rural income changes are nevertheless more sensitive to model specification. When agricultural supply is inelastic, loosening agricultural supply constraints results in large increases in production and household incomes. More generally, as in the CGE model, technological changes that increase agricultural productivity generate important real-income linkages in rural farm and non-farm economies. Even though households do not benefit directly from these supply increases, their total incomes rise substantially; village demands for non-farm goods from outside markets increase by a similar magnitude. When prices are endogenous, the terms of trade turn sharply against production sectors experiencing productivity gains. Lower prices for food mitigate the positive income effects of technological change for rural producers, but generate real-income gains for consumers.

In short, where technological and other constraints result in inelastic agricultural supply, measures to increase supply response are critical for strengthening farm/non-farm linkages associated with rural household incomes. Loosening agricultural supply constraints in itself stimulates the non-farm economy by creating rural income multipliers and raises the real incomes of rural and urban consumers.

Further research is needed to inform the design of policies to promote the RNFE. The present study, by constructing village and microregional general equilibrium models represents a step in the direction of better understanding of rural farm/non-farm linkages. In the light of the rapid growth of intermediate cities in rural areas of LDCs, an extension of this research to model linkages between villages or microregions and intermediate cities would constitute an appropriate next step.

REFERENCES

Adelman, I. 1984. Beyond export-led growth. *World Development*, 12(9): 937–949.

Adelman, I., Taylor, J.E. & Vogel, S. 1988. Life in a Mexican village: a SAM perspective. *Journal of Development Studies,* 25(5): 24.

Barnum, H.N. & Squire, L. 1979. An econometric application of the theory of the farm household. *Journal of Development Economics*, 6: 79–102.

Bardhan, P. 1988. Alternative approaches to development economics. In H. Chenery & T. N. Srinivasan, eds., *Handbook of development economics*, Vol. 1. New York, USA, Elsevier Science Publishers.

Braverman, A. & Hammer, J.S. 1986. Multimarket analysis of agricultural pricing policies in Senegal. In I. Singh, L. Squire & J. Strauss, J., eds., *Agricultural household models, extensions, applications and policy*, pp. 233–254.

Bellon, M. & Taylor, J.E. 1993. "Folk" soil taxonomy and the partial adoption of new seed varieties. *Economic Development and Cultural Change*, 41(4): 763–786.

De Janvry, A., Fafchamps, M. & Sadoulet, E. 1991. Peasant household behaviour with missing markets: some paradoxes explained. *The Economic Journal*, 101:1 400–1 417.

Feder, G., Just, R.E. & Zilberman, D. 1985. Adoption of agricultural innovations in developing countries: a survey. *Economic Development and Cultural Change*, 3: 255–298.

Golan, E.H. 1990. *Land tenure reform in Senegal: an economic study from the peanut basin*. Madison, Wis., USA, University of Wisconsin. (Land Tenure Center Research Paper No. 101.)

Haggblade, S., Hammer, J.S. & Hazell, P. 1991. Modelling agricultural growth multipliers. *American Journal of Agricultural Economics* (May):361–374.

Hart, G. 1989. The growth linkages controversy: some lessons from the Muda case. *Journal of Development Studies*, 25(4): 571–575.

Hazell, P.B.S. & Roell, A. 1983. *Rural growth linkages: household expenditure patterns in Malaysia and Nigeria*. Washington DC, IFPRI. (Research Report No. 41.)

Hirschman, A.O. 1958. *The strategy of economic development*. New Haven, Conn., USA, Yale University Press.

Lewis, B.D. & Thorbecke, E. 1992. District-level economic linkages in Kenya: evidence based on a small regional social accounting matrix. *World Development*, 20(6): 881–897.

Lopez, R.E. 1986. Structural models of the farm household that allow for interdependent utility and profit-maximization decisions. In I. Singh, L. Squire & J. Strauss, eds., *Agricultural household models, extensions, applications and policy*, pp. 306–326.

Mellor, J. 1976. *The new economics of growth*. Ithaca, NY, Cornell University Press.

Meng, E.T. 1997. *Household varietal choice decisions and policy implications for the conservation of wheat landraces in Turkey*. Texcoco, Mexico, CIMMYT.

Otsuka, K. & Reardon, T. 1998. *Lessons from rural industrialization in East Asia: are they applicable to Africa?* Washington DC, IFPRI/World Bank. (Paper for IFPRI/World Bank conference Strategies for Stimulating Growth of the Rural Non-farm Economy in Developing Countries, 17–21 May, Airlie House, Va., USA.)

Parikh, A. & Thorbecke, E. 1996. Impact of rural industrialization on village life and economy: a SAM approach. *Economic Development and Cultural Change*, 44(2): 351–77.

Ralston, K. 1992. *An economic analysis of factors affecting nutritional status of households in rural west Java, Indonesia.* Berkeley, Calif., USA, University of California (Ph.D. thesis).

Rangarajan, V. 1982. *Agricultural growth and industrial performance in India.* Washington DC, IFPRI. (Research Report No. 33.)

Reardon, T., Delgado, C. & Matlon, P. 1992. Determinants and effects of income diversification amongst farm households in Burkina Faso. *Journal of Development Studies*, 28(2): 264–296.

Singh, I., Squire, L. & Strauss, J. 1986. *Agricultural household models, extensions, applications and policy.* Washington DC, World Bank, Johns Hopkins University Press.

Smale, M., Just, R. & Leathers, H. 1994. Land allocation in HYV adoption models: an investigation of alternative explanations. *American Journal of Agricultural Economics*, 76(3): 535–546.

Subramanian, S. & Sadoulet, E. 1990. The transmission of production fluctuations and technical change in a village economy: a social accounting matrix approach. *Economic Development and Cultural Change*, 39(1): 131–173.

Taylor, J.E. 1995. *Microeconomy-wide models for migration and policy analysis: an application to rural Mexico.* Paris, OECD.

Taylor, J.E. & Adelman, I. 1996. *Village economies: the design, estimation and use of village-wide economic models.* Cambridge, UK, CUP.

Yúnez-Naude, A., Barceinas, F. & Taylor, J.E. 1994. Reflexiones sobre la biodiversidad genética de las emillas; problemas de análisis y el caso del maíz en México. In A. Yúnez-Naude, ed., *Medio ambiente: problemas y soluciones.* Mexico City, El Colegio de México.

Chapter 3
The rural non-farm economy and farm/non-farm linkages in Querétaro, Mexico

Fernando Rello and Marcel Morales

INTRODUCTION

The objective of this chapter is to improve understanding of the Mexican non-farm rural economy. This objective reflects a growing perception among Mexican researchers and government officials that the rural economy goes far beyond agriculture. Studies from several countries, including Mexico, have shown that sources of employment and income for rural families are varied and that agriculture is no longer the primary occupation or income source of most rural inhabitants. This suggests a need to improve understanding of the structure and dynamics of the non-farm rural economy and its links to the farm sector.

To study the non-farm rural economy adequately, it is necessary to examine the intersectoral linkages established in particular regions and their capacity for generating local employment. Such a study calls for the incorporation of a spatial dimension, and close analysis of the role of intermediate cities and medium and small rural towns in creating and propagating these linkages. Fuller understanding of the linkages and of the dynamics of industrial development in a particular regional and institutional context provides a useful basis for identifying sectoral policies; the Querétaro region was chosen as a case study with this in mind. Querétaro is a dynamic intermediate city that has close links with its hinterland; it lies in the centre of Mexico and is linked with the country's economic centre, Mexico City. This makes it an ideal location for examining linkages between the farm and non-farm rural economy.

The remainder of this chapter is divided into three sections. The second section provides information on the state of Querétaro, with a description of the geographic and economic characteristics of the region and an analysis of the main characteristics of the agricultural sector, including sources of employment and income for rural families. In the third section, several case studies are presented that examine the variety of linkages between the farm and non-farm sectors and the employment generated by these linkages. In the fourth section, observations are made regarding sector policies directed at rural regions and regional institutions; recommendations for policy intervention are offered.

THE QUERÉTARO REGION

Description of the region

The state of Querétaro is situated in central Mexico. It is divided into 18 municipalities, the most important of which are the intermediate cities of Querétaro and San Juan del Rio, which are population and economic centres. The Querétaro region is the most important region in the state, because it is the centre of economic activity and has high-quality soil suitable for arable and cattle farming (COEPO, 1995). The Mexico–Querétaro highway crosses the region, connecting it with the largest national market, Mexico City, the north of the country and the United States. The region is thus an almost obligatory step in commercial flow between the north of the country and Mexico City. This geographical advantage led to the establishment of important industries in Querétaro, mainly during the 1970s.

In 1980, the total population of the State of Querétaro was 739 605 inhabitants; the growth rate was 4.2 percent during 1970–1980. Between 1980 and 1990 the population growth rate dropped to 3.6 percent; the population had reached 1 051 235 inhabitants by 1990. Table 1 provides a breakdown of the population by size of locality. In 1990, four localities were considered urban and included 46.6 percent of the state population; the city of Querétaro accounted for 36.7 percent of the state population. The population concentration reaches 42.6 percent if it is grouped with San Juan del Rio, the other locality that has more than 50 000 inhabitants. There were 1 420 localities with populations ranging from 1 to 1 999 inhabitants, comprising 37.2 percent of the total population. Of the remaining population, 16 percent are in localities with between 2 000 and 14 999 inhabitants; 4 percent are in the medium towns of Tequisquiapan and El Pueblito, which each have 15 000–25 000 inhabitants. The total population is thus spatially distributed in a manner that is both dispersed and concentrated, reproducing at state level the phenomenon prevailing at national level, where there is only one primary city with substantial economic activity and population.

For the majority of the population in the region, living standards are not high, especially in rural municipalities where levels of employment, health, education and nutrition are low. The standard of living diminishes in proportion to the distance from the Querétaro–San Juan del Rio industrial corridor. At

TABLE 1
Querétaro: population by size of locality, 1990

Size of locality (inhabitants)	Population	%
Total	1 051 235	100.00
1–99	21 568	2.05
100–499	126 594	12.04
500–999	100 149	9.53
1 000–1 999	143 009	13.60
2 000–2 499	32 076	3.05
2 500–4 999	77 844	7.41
5 000–9 999	48 789	4.64
10 000–14 999	11 798	1.12
15 000–19 999	19 231	1.83
20 000–49 999	23 022	2.19
50 000–99 999	61 652	5.86
100 000–499 999	385 503	36.67

Source: Querétaro XI Censo General de Población y Vivienda, 1990. Instituto Nacional de Estadística Geografía e Informática (INEGI).

national level, Querétaro was considered to be fairly marginalized by 1990, occupying 14th place out of the 31 Mexican states (CONAPO, 1993).

Infrastructure development in the region is concentrated in the municipalities that form the Querétaro–San Juan del Rio corridor. The other main roads in the state link the municipal capitals to the cities of Querétaro and San Juan del Rio. The most important highway axis is thus the section from San Juan del Rio to Querétaro, on the Mexico–Querétaro toll road. It must be emphasized that although this highway links Mexico City with the north and the west, approximately 55 percent of the traffic circulating through this corridor begins or terminates within the state; the corridor does not function only as a thoroughfare (Martner, 1991). Most of the irrigation and water-storage areas are also concentrated in the Querétaro and San Juan del Rio municipalities.

Economic activities

The structure of economic activity in Querétaro is closely connected to the transformations occurring at the national level. The sectoral contribution of the state GDP is characterized by a decline in importance of primary activities – that is, agriculture – and an increase in secondary activities – manufacturing – and tertiary activities – services. In 1970, the primary sector contributed

18.0 percent to state GDP, the secondary sector 36.9 percent and the tertiary sector 45.2 percent. By 1993, the primary sector had declined to only 4.6 percent of state GDP, while the secondary sector maintained its contribution with 36.2 percent and the tertiary sector grew significantly to 59.3 percent. In the primary sector, there has been a decrease in the contribution of crop production and an increase in the importance of animal production. By 1988, crop production accounted for 34.0 percent of agricultural activity; by 1993 it had fallen to 25.8 percent (INEGI, 1996).

In terms of employment generation, in 1990, the tertiary sector generated most employment, with 41.8 percent of the total, followed by the secondary sector with 37.0 percent and the primary sector with 17.9 percent. It should be noted that although the primary sector directly generates only 4.6 percent of the state GDP, it employs a substantially larger proportion of the population. At municipal level, there are important differences in economic activity. The smaller

TABLE 2

Main features of manufacturing firms per municipality, 1993

Municipality	No. of firms	Average total of employees
Querétaro	**3 056**	**60 518**
Amealco	26	103
Pinal de Amoles	–	13
Arroyo seco	–	10
Cadereyta	91	1 085
Colón	12	85
Corregidora	128	3 392
Ezequiel M.	299	1 998
Huimilpan	15	64
Jalpan de S.	20	75
Landa de M.	–	1
Marqués, el	64	1 538
Pedro Escobedo	45	1 341
Peñamiller	–	21
Querétaro	1 537	33 036
San Joaquín	7	19
San Juan del R.	437	16 271
Tequisquiapan	340	1 330
Toliman	15	136

Source: Base de datos municipal, SIMBAD. INEGI.

TABLE 3
Main features of manufacturing firms per economic activity, 1993

Economic activity	No. of firms	%	No. of employees	%
Querétaro	3 054	100.0	60 518	100.0
Food, beverages and tobacco	839	27.5	10 440	17.3
Textiles, garments and leather	400	13.1	10 015	16.5
Wood goods	381	12.5	1 449	2.4
Paper and editorial activity	194	6.4	4 314	7.1
Chemicals, oil, coal	100	3.3	7 197	11.9
Plastic and oilcloth goods				
Mineral, non-metallic goods	258	8.4	3 098	5.1
Basic metal Industry	–	–	122	0.2
Equipment and machinery	852	27.9	23 668	39.1
Other manufacturing industries	–	–	215	0.4

Source: XIV Censo Industrial. Censos Económicos, 1994. INEGI.

municipalities, including Amealco, Colón, Huimilpan and Marqués, are more dependent on primary activities; the larger municipalities depend on secondary and tertiary activities. As can be seen in Table 2, the location of industrial activity is highly concentrated in a few municipalities, with 75 percent of manufacturing establishments located in three municipalities: Querétaro with 50.3 percent, San Juan del Rio with 14.3 percent and Tequisquiapan with 11.1 percent.

Table 3 shows further information on Querétaro manufacturing firms by subsector. In 1993, 3 054 firms had been established in Querétaro, providing employment for 60 518 workers. Measured in terms of employment generation, the most important subsectors are equipment and machinery, followed by food, beverages and tobacco and textiles, garments and leather. The first of these is likely to be linked to agriculture through farmers' purchases of machinery, the latter two by the supply of inputs from the farm sector.

Agriculture in the region

A description of the basic agricultural characteristics of the region is given here in order to put rural producers in context. First, the region includes the municipalities of Amealco, Colon, Huimilpan and El Marques, which together cover 149 500 ha, or 23 percent of the state of Querétaro. Of this area, more than half is cultivated land situated in an extensive valley; 60 percent of the cultivated land is irrigated. Animal production is an important activity, particularly poultry, meat and milk production, in that order. A number of farms and ranches have been mechanized and have made heavy investments in production of animal products. There is also substantial home-based animal production that brings in a complementary income to numerous farm families. Crop and animal production are complementary for these households, because fodder such as sorghum, alfalfa and oats are cultivated for feed, and maize stubble is used to feed home-reared animals.

Table 4 presents information on the crops cultivated in the region. The primary crop is clearly maize, which is either grown alone or intercropped with beans. The crop pattern has been modified over time, according to the evolution of the profitability of different agricultural products, influenced by the market and policy. No historical time series are available of the evolution of the cultivated

TABLE 4

Main crops per municipality (cultivated land in ha, 1995–1996)

	Municipality			
	Amealco	**El Marques**	**Huimilpan**	**Colon**
Cultivated land	22 547	17 907	8 779	14 562
Maize	2 547	9 825	7 472	3 813
%	100	54.90	85.10	26.20
Intercalated maize *	–	4 297	840	6 578
%		24.00	9.60	45.20
Sorghum	–	262	–	1 550
%		1.50		10.60
Beans	–	878	273	858
%		4.90	3.10	5.90
Barley	–	600	27	800
%		3.30	0.30	5.50
Wheat	–	180	104	105
%		1.00	1.20	0.70
Chick peas (fodder)	–	–	–	–
%				
Oats (fodder)	–	130	5	190
%		0.70	0.10	1.30
Alfalfa	–	1 735	58	668
%		9.70	0.60	4.60

* Maize intercalated with beans and other crops.
Source: Querétaro VII Censo Agricola-Ganadero. INEGI.

surface corresponding to the municipalities under study, but there are the agricultural time series for the state of Querétaro for the period 1988–1997. It should be noted that there have been changes in the use of crops; the most notable features of this have been the rise in maize production, although this has been partially reversed in recent years, and the decline in sorghum production. The land area under maize increased because of attractive sale prices, favoured by policy, and because of an increase in maize productivity in irrigated lands resulting from the application of a modern technological package. Barley is another expanding crop with increases resulting from a rise in demand fuelled by the beer industry and rises in productivity. The most dynamic products during the last five years have been vegetable crops.

Characteristics of rural households

One objective of this research is to develop a typology of households as a means of identifying the structure and functioning of rural production units. This helps to understand the types of household that are likely to be linked to the non-farm rural economy and the types of linkages that occur. Traditionally, typologies of rural producers have been based on agricultural land ownership. This criterion is insufficient and partial in the present circumstances, however, where only a part of the rural household income comes from land cultivation. This is particularly true for land-poor households for whom agricultural income is a small and constantly diminishing part of their total income. Forming a typology thus requires consideration of the structure of employment and sources of income among rural households. Sources of household income are varied. Several methods can be used to develop a typology. One of these could consist of identifying family sources of income and classifying these according to the relative importance of each source. Another method consists of classifying them according to the availability of natural, human and social capital that provides them with an income from several sources. In this section, an intermediate approach has been taken: households have been classified first according to their natural capital in terms of quantity of land in possession, and then evaluated based on their agricultural practices and sources of income.

To gather appropriate information to classify and evaluate household types, a survey of 285 households in 40 rural communities in the study region was conducted. Households were distributed across the study region as follows: 47 percent in the municipality of Amealco, 20 percent in Huimilpan, 10 percent in Colon, 14 percent in Marques and the remainder distributed between Cadereyta and San Juan del Rio. The survey questionnaire included information on productive resources, technological level, commercialization, access to services,

TABLE 5

Typology of producers

Type	Land owned (ha)	Number of producers	% of producers	% of cultivated land
0	0	12	4.2	0
1	< 3	80	28.0	9.8
2	3–5	71	24.9	20.0
3	5–10	100	35.1	44.8
4	> 10	22	7.7	25.4
Total		285	100.0	100.0

Source: Household survey.

occupations of household members and income sources. The stratification of households was based on land ownership, because the portfolio of productive activities depends on agriculture.

A first differentiation of households was made between those that own land and those that do not own land but have access to land through the rental market. The latter constitute the "zero stratum" and represent 4.2 percent of the sample. Within the set of households who own land, there are four strata. The distribution of the 285 producers across strata is shown in Table 5, which indicates that the medium landowner households with 5–10 ha, accounting for 35.1 percent, are the most abundant and own the greatest total proportion of land. There is a smallholder sector that makes up 28 percent of households and 9.8 percent of land with an average of 1.9 ha per household; it coexists with a stratum of large households, 7.7 percent of households and 25.4 percent of land, reflecting the unequal distribution of land prevailing in the rest of Mexico. The average property of this latter group is 18.3 ha per capita, however, which does not represent an high level of concentration of land ownership.

Table 6 shows the characteristics of each household type. With regard to land tenure, the region is eminently *ejidal*, because more than two-thirds of the land area belongs to this property regime: 88 percent of households possess *ejidal* lands and 24 percent possess private properties, which suggests that 12 percent of producers combine *ejidal* tenure with private property. Leasing land is a common practice for 15 percent of households; this frequency has significant variations, however, if the behaviour of each one of the strata is examined. Of non-owner households, 83 percent use the land market to gain an average of 12.4 ha of land. Renting land is also common among households

TABLE 6
Typology of households and sources of income

	Type 0	Type 1	Type 2	Type 3	Type 4
Agriculture					
Owned land (ha)	0	1.9	4.5	7.1	18.3
Rented land (ha)	12.4	2.0	2.7	3.2	7.4
Farmers using a tractor (%)	58	41	45	68	91
Farmers using hybrids seeds (%)	50	13	21	43	59
Farmers using fertilisers (%)			80		95
Cattle owned (head)	40	6.4	7.1	8.4	8.0
Households owning cattle (%)	34	35	49	57	82
Income sources					
Average family income (pesos)	21 620	20 055	18 814	26 144	58 982
Crop income (%)	52.1	34.2	46.0	52.7	42.6
Animal production income (%)	17.7	15.2	15.1	21.9	14.9
Family industry income (%)	0	0	1	0.4	4.6
Ag. wage labour (%)	0.8	6.8	9.4	9.2	0.6
Artisanal and crafts income (%)	2.8	13.4	3.7	3.7	4.0
Commercial income (%)	9.3	10.2	15.3	4.8	21
Remittances (%)	1	3	0.2	2.3	4.4
Salaries (%)	17.3	17.3	9.3	4.9	7.8

Source: Household survey.

who already have a relatively large amount of land (stratum 4), 32 percent of whom lease an average of 7.4 ha per farmer. The intermediate strata are less likely to rent land, and rent less land. The level of technological advancement is closely related to land ownership: greater land ownership appears to result in more intensive use of modern inputs and machinery. It is interesting to note that the renting producers (stratum 0) have a technological level superior to the average, which indicates that they invest capital to increase the value of the rented land.

In spite of the fact that only 15 percent of the households are tractor owners, there is considerable use of tractors among households of several strata; 56 percent of the households use tractors for preparation of their land. This frequency varies from 41 percent for stratum 1 to 91 percent for stratum 4. Thirty-one percent of households rely on the use of hybrid seed; the evidence suggests a strong positive correlation between the amount of land used, including rented land, and use of hybrid seed.

A high percentage of households own animals and are involved in animal production as a primary or complementary activity. Ownership of animals increased in the period 1992–1997, when the economic crisis in the rural sector worsened, possibly since it was viewed as a safe means of accumulating wealth and an insurance against problems. A number of different types of animals are owned, primarily chickens and cows, although rearing of pigs and sheep is also notable.

Data on household income generation indicates that sources of employment and income vary substantially across households, so much so that virtually no household lives entirely off primary activities. Although practically all households declare their agricultural activities as a priority activity, only about half are crop production and 15–20 percent are animal production. The remaining income comes from a variety of other activities including day labour, commercial activities and salaried employment. The diversity across household types results largely from differences in these off-farm activities. Type 0 households are more likely to be salaried employees or involved in commercial activities. Type 1 households with limited land tend to produce arts and crafts for the market, to have another commercial activity or to be salaried employees. Type 2 and type 3 households obtain income through agricultural day labour; type 2 households are also involved in commercial activities. Type 4 large landholders are involved in a number of off-farm activities, particularly commercial activities.

For all types of producers, off-farm sources of income are important, even for the producers who have more than 10 ha, who in theory could live from agriculture and cattle rearing. This indicates that participation in several sources of employment and agro-industrial and commercial activities forms part of the economic strategies of rural households. The corollary to this observation is that all policies designed to raise employment and rural incomes would tend to further development of productive links and improve rural family members' participation in different regional markets.

AGRO-INDUSTRY ORGANIZATION AND LINKAGES: CASE STUDIES FROM QUERÉTARO

Evaluating the linkages between the farm and non-farm sectors of the rural economy can be conducted from a macroeconomic perspective, using an input-output matrix, as in Chapter 2, or through case studies, using data collected from surveys and interviews. This section follows the latter approach using information collected from the field on six different agro-industrial systems: maize, cempasúchitl, frozen and processed vegetables, barley and beer, poultry and milk. A case study approach is taken because farm/non-farm linkages tend to be specific to certain agro-industrial systems, and a case study approach is better suited to understanding the details of these systems. This section is divided according to the agro-industrial systems in the region, in each of which the important linkages between producers, agro-industry and the local economy are highlighted. Economic agents that have played a role in forming linkages are discussed as necessary. The data is based on surveys of agricultural producers, as discussed earlier, and extensive field interviews with informants in primary agro-industrial firms and public officials operating in each system.

Maize production system

Maize is the primary agricultural product in Querétaro. It is the basic staple crop for families in the region, an important commercial crop and a major source of fodder for animal production. The Ministry of Agriculture estimates that maize for food and feed represents as much as 30 percent of the total harvest. Although central to household consumption, maize is complemented by other foods including beans, fruits, vegetables, poultry, milk and beef, all of which are produced locally on family plots or on common lands, or imported from local towns and intermediate cities. To speak about a maize system is thus an oversimplification of a complex food-production system of which maize is an integral part. Many of the linkages described in this subsection are the same for a number of other crops; for this reason, this crop production system is discussed in greater detail than the other systems.

Two significant events have occurred over the past few decades that have had an important influence on the maize-production system. First, there have been dramatic changes in the technology and agricultural practices used by farmers. There has been an ongoing transition from a traditional agricultural system based on locally grown seeds, organic fertilizer, animal traction, rainfed corn and traditional knowledge, to a modern agricultural system based on hybrid seeds, chemical fertilizers and pesticides, tractors, irrigated water and technical assistance. This has had a profound effect on the linkages within the maize-production system and far-reaching consequences for the local non-farm sector. Second, the role of the state in Mexico has changed and with it there have been changes in macroeconomic conditions. As in many developing countries, the government has historically played a very active role in agricultural production systems, particularly through the Ministry of Agriculture. Maize was particularly targeted as the primary staple crop in Mexico, so scaling-down of the Ministry of Agriculture and elimination of programmes during 1989–1996 hit maize producers more severely than producers of other crops. These changes occurred at precisely the same time as a difficult transition from rainfed to irrigated maize cultivation, caused by the crisis in sorghum, the main irrigated crop in the region. Liberalization led to the entry of very cheap imported sorghum; internal prices for sorghum collapsed to a point where cultivation was no longer profitable. The shift from sorghum to maize was inevitable, but farmers had not mastered the appropriate technology for cultivating irrigated maize. Over-fertilization, application of too many agrochemicals and wastage of water left yields well below potential and caused problems of salinization and deterioration of the natural productivity of soil. Lack of appropriate technical assistance has been a serious limitation to potential increases in productivity and local incomes.

Against this changing context, the linkages within the maize production system can now be considered. During the 1980s and into the 1990s, the impetus for changes in technological and agricultural practices came from the Ministry of Agriculture and its army of agronomists. At that time, the Mexican state was exceptionally paternalistic and interventionist in the rural sector, using various public institutions and state-owned companies to manage agricultural production. The main public agents influencing the maize-production chain were FERTIMEX, a parastatal with a monopoly on production and importation of chemical fertilizers, BANRURAL, a public agricultural bank that provided the greater part of credit available to small and medium farmers, CONASUPO, a parastatal that purchased the bulk of the maize harvest at fixed prices, and the Ministry of Agriculture, which provided subsidized technical assistance to farmers. The Ministry of Agriculture designed and implemented the agricultural modernization programmes in the region, coordinated other institutions, distributed fertilizer directly to farmers' organizations, administered fertilizer deposits and ran demonstration and experimental plots. Subsidized credit for maize production was available; CONASUPO represented a secure market outlet. The state dominated the production chain and was linked to farmers through a number of channels.

Throughout this period, demand for fertilizer grew steadily in the region, even after the disappearance of the fertilizer programme. According to the Ministry of Agriculture, half the area devoted to maize was fertilized in 1996; all irrigated maize was fertilized and about a third of rainfed maize used this input. FERTIMEX and its distribution network initially satisfied this increase in demand, but as a result of the financial crisis and policy changes noted above,

FERTIMEX was dismantled and its installations and equipment sold to private firms and farmers' organizations. This created new opportunities for formation and growth of private firms and farmers' organizations; several fertilizer-distribution firms are now operating in the region.

Interviews with managers of two of these firms revealed interesting information regarding input flows and job creation. Fertilizers are manufactured in large industrial centres outside the region and imported by distribution firms, so there are no direct backward production linkages. Distribution firms are medium-sized and employ an average of ten permanent employees; they also create a demand for transportation services from local firms and contribute to the creation of jobs for truck drivers and mechanics. Their global impact on job creation is not substantial, but it is significant at local level because they are located in small and medium-sized rural towns.

Hybrid seeds are another major component of the technological change that took place in the region: about 30 percent of land under maize is currently planted with hybrid seeds. Expanded use of hybrid seeds occurred during the 1990s, however, a decade after the fertilizer boom at a moment when public technical assistance had declined dramatically. This means that although the public sector played an important role in the dissemination of fertilizers through provision of technical assistance, it had only a minor role in the expansion of hybrid-seed use. Private, medium-sized commercial firms play the main role in introducing hybrid seeds.

Hybrids cannot be grown locally, so they must be bought at the beginning of each agricultural cycle. The continuous and growing demand for seeds is met by agrochemical shops, which distribute seeds produced by large international companies such as Asgrow and Pioneer. Agrochemical and seed shops are a source of employment, especially for skilled labour, because one of their most important functions is to provide advice and technical assistance to customers. Semillas y Agroquímicos San Jorge, for example, a company that serves approximately 3 000 customers and that has subsidiaries in other regions, employs six agronomists and six administrative and sales personnel.

In addition to fertilizers and hybrid seeds, a number of other inputs are linked to maize production. To understand these backward linkages, household heads were questioned as part of the household survey about their primary places of purchase of inputs and machinery, hire and repair of machinery and hire of work animals. Based on these results and other information gathered from the region, the following conclusions can be drawn.

1. Farmers tend to buy substantial amounts of seed and animal feed and substantial numbers of draught animals in the local area. Services, including repair services and hire of tractors and estate cars, are often carried out in the local area. Only 15 percent of respondents claimed to be tractor owners, for example, but more than 55 percent claimed regular tractor use, which suggests that there is a market for hired tractors. The same is true for trucks and estate cars. The production of maize and other crops generates substantial local linkages.

2. Fertilizers, agrochemicals such as insecticides and pesticides, tools and animal feed are purchased in specialized shops located in medium-sized towns, which also provide repair services for tractors and estate cars. Results indicate that medium-sized towns have significant links to agriculture.

3. Querétaro is the main source of major purchases such as tractors (63 percent), harvesters (73 percent) and estate cars (45 percent). Authorized dealers of major machinery brands such as Ford and John Deere are located there. Querétaro is also a main provider of other specialized machinery such as electric pumps.

4. In addition to local dealers of draught animals, regional fairs are an important commercial channel for cattle, draught animals and other inputs. The fairs generally involve substantial local participation and provide a mechanism for linking commercial interests.

In addition to the range of production inputs that provide backward linkages to agricultural production, credit and insurance play an important role in fostering linkages between sectors; such companies directly employ people in the financial sector. Credit for agriculture has fallen dramatically during the structural-adjustment programmes of recent decades. BANRURAL, for instance, used to finance 40 000 ha in Rural Development District 004, which overlaps most of the selected region, but now finances only 3 000 ha. Agricultural insurance coverage by the official insurance company AGROASEMEX has also been reduced substantially. AGROASEMEX prefers not to cover rainfed agriculture; when it does so, its prices are so high – more than 25 percent of the cost of production – that farmers cannot afford insurance. Limited access to agricultural credit and insurance are major problems that may limit backward linkages.

Commercialization of maize is the most important of the forward production linkages from the farm, with approximately 70 percent of the harvest sold in local markets. Households use the remaining 30 percent for food, fodder or seed. During the 1980s, CONASUPO was responsible for purchasing unlimited quantities of maize at a fixed price determined at the beginning of each agricultural cycle, and would generally buy more than 80 percent of the harvest sold. This volume of purchase, made possible by a chain of storehouses located within the region, permitted CONASUPO to regulate the regional price of corn. Scaling-down and elimination of administered prices changed the commercialization pattern drastically. CONASUPO is still buying maize, but to a lesser extent than before, and is now competing with private local and regional grain merchants and with large private industrial companies, which use corn as an input in processing. Importation of foreign corn is increasing and local prices tend to follow international market prices.

A small portion of the maize harvest is exchanged in local shops for food and other consumption goods, but most is sold in the local market. The primary buyers are local and regional grain merchants and CONASUPO, although market shares vary. Information on the regional maize market from the survey of rural households indicates that the largest maize purchaser is private merchants (73 percent) followed by CONASUPO (25 percent). Direct purchases of maize from farmers by agro-industries are still very low, although regional merchants are presumably acting as intermediaries and selling to these companies. Prices paid by merchants are slightly lower than those paid by CONASUPO, but merchants pay immediately and go directly to plots where they collect and transport the harvest, with no extra cost to producers.

In terms of employment generation, maize is cultivated almost entirely with family labour; contracted labour is almost non-existent. When labour needs exceed family labour during certain peak periods, labour is exchanged through traditional mutual help systems. Hired labour is more common in the cultivation of commercial crops than maize and other basic crops. According to this survey, the total workdays of hired labour used in the cultivation of the four main crops by all households interviewed sum to only 3 661; divided by 250, a standard workday per annum for a fixed employee, gives a total of 14.6 workers. This means that in their agricultural activities during 1996, the 286 producing households that were interviewed generated an equivalent of only 14.6 permanent jobs.

Private grain-processing companies, NGOs and farmers' associations have recently played an important role in fostering linkages within the maize system. MASECA, for instance, a large company that purchases significant quantities of maize to produce corn flour for tortillas, has created an association called the Corn Club, linking the different interests that work in the maize market. MASECA needs a constant and secure flow of high-quality white corn. To assure this flow, it contracts with farmers for a specific number of hectares of cultivated maize.

Farmers promise to sell the total harvest to MASECA, and the firm promises to buy the harvest at a specified price.

Production of good-quality white corn, as specified in contracts, requires cultivation of hybrid seeds and the use of a particular technological package; irrigation is also stipulated as an obligation on the part of the producer. To finance this package and avoid any bottleneck caused by credit constraints, the Fideicomisos Instituidos en Relación con la Agricultura (FIRA) trustship of the Central Bank (Banco de México) was established to foster rural development and open credit lines for farmers belonging to the Corn Club. Credit provision is handled by a private bank participating in the scheme, not FIRA, which gives credit directly to farmers at low but not subsidized interest rates and provides agricultural insurance. At the end of the harvest, total production is delivered to MASECA and the firm pays credit and interest back to the bank. As a final step, MASECA pays farmers the value of production minus the credit at the fixed price in the contract. By means of this association scheme, positive linkages are created and the credit, insurance and commercialization problems are solved to the benefit of all participants.

Farmers associations and NGOs also foster and reinforce positive linkages, as in the case of the Unión de Ejidos Francisco Villa. This association was formed by several regional *ejidos* in 1986; its membership amounts to 2 366 *ejidatarios*. The association's primary activity is purchase of large quantities of fertilizer for distribution to members at low prices. Before the scaling-down of FERTIMEX, the union had managed to get a concession from the state-owned company for local distribution of fertilizer. When FERTIMEX disappeared and the fertilizer market was privatized, the Unión de Ejidos, fearing price increases of this vital input, strengthened its role as distributor in competition with with new entrants into the market. The association also entered into an alliance with Comercializadora de Occidente (COMAGRO), a large social enterprise whose main aim is to deliver inputs and services to its members. COMAGRO's scale of operation allows it to purchase enormous quantities of fertilizer, herbicides and pesticides, and to negotiate favourable prices with large private companies, including multinationals. COMAGRO sells fertilizers to the Unión de Ejidos at low prices, which enables it to compete successfully with other local distributors in the Querétaro region. In the process, COMAGRO provides its members with technical assistance and helps to find finance and market outlets; for example, it signed a significant corn-purchase contract with MASECA on behalf of its members. In this way, COMAGRO contributes to the development of linkages favourable to producers and integrates them with other economic agents.

The Unión de Ejidos Francisco Villa has gone into the insurance business in association with AGROASEMEC, the public agricultural insurance company. When the previous government insurance enterprise ANAGSA was dismantled, the union decided to organize an insurance scheme to protect its members, particularly those with irrigated land and high-cost technological packages. The union's insurance scheme covers only capital invested in the application of a technological package recommended by a national institute of agricultural research, INIFAP. In case of harvest losses, the union's insurance scheme and AGROASEMEC cover the costs; the union is, in other words, a sort of intermediary between AGROASEMEC and the farmers.

Cempasúchitl production

Cempasúchitl is an intensely coloured yellow flower that is transformed into flour and used as a pigment by the poultry industry. When added to animal feed, cempasúchitl flour causes chicken skin to acquire a golden colour and egg yolks to be more orange, both of which are attractive to consumers. Chicken consumption has been increasing in Mexico as a result of rising incomes and favourable chicken prices relative to beef and pork. Demand for natural pigments, among other inputs used by the poultry industry, has been growing steadily. Pigments are not produced by the poultry industry itself but by specialized firms, which need cempasúchitl flower and red chilli as basic inputs. In order to ensure a constant flow of input, these firms offer attractive prices and technical assistance to peasants willing to initiate cempasúchitl cultivation.

Spurred by increasing demand for the flower, a number of producers have shifted from maize production to the labour-intensive cempasúchitl production. Total production costs for cempasúchitl are 2.5 times greater than for maize primarily because it is more labour-intensive and because more intensive application of fertilizers and insecticides is required. One hectare of maize can be cultivated using 32 workdays during a six-month cycle; one hectare of cempasúchitl requires 139 workdays in a four-month cycle. Based on average yields of 3 mt/ha for maize and 15 mt/ha for cempasúchitl, maize gives profits of 895 pesos per ha/cycle, while cempasúchitl gives profits of 5 865 pesos per ha/cycle – a significant difference.

Entry into cempasúchitl production is not easy, because it requires irrigation and has high input costs. Credit available from BANRURAL, the main financial source in the region, amounts to only 50 percent of the cost of production, so the remaining production costs have to be financed using savings or other income. This inhibits production by the poorer strata of farming households; poor households may benefit from employment generated by increased production, however. In Santiago Mezquititlan, a low-income region near Amealco, for instance, 600 ha were cultivated in the 1998 cycle, generating a need for 83 400 workdays in four months; most of this labour requirement was met by wage labour from poorer households.

There are two pigment-producing agribusinesses operating in the region, Alcosa and Bioquimex, both of which are located in intermediate cities. Based on interviews of Alcosa staff, the following information regarding linkages was gathered. Alcosa is a medium-sized national firm, with four plants in central and northwest Mexico. The regional plant in Apaseo El Grande, in the vicinity of Celaya, consumes 14 000 tons of flowers each year, which come from some 300 small producers in Santiago, Mezquititlan. A small quantity of flour is imported from India. Contract farming is used in all cases. Contracts specify planting time, quantities to be purchased, prices to be paid, delivery sites, quality of input, supervision, monitoring and penalties. The company does not give credit, but provides flower seeds and offers technical assistance, including how to sow the seeds, prepare the soil, apply fertilizers and insecticides and when to cut the flower.

Alcosa and Bioquimex prefer contract farming to imports so that they can maintain flower quality standards. In the past, the companies relied more heavily on imports, but quality was unsatisfactory. Now they are trying to establish a network of local producers and are increasingly using contract farming. This is in contrast to the case of red chillies, which have a well established national market where both firms purchase their required quantities at prevailing prices on the spot market. There is therefore no need for contract farming. According to Alcosa managers, chilli is a risky crop and farmers are reluctant to sign contracts for it.

The primary local effects of pigment production are through production linkages from firms to farmers, and employment linkages from farmers to labourers. Direct linkages in the intermediate cities are few, but not negligible. Alcosa employs 72 permanent technical, administrative and manual workers who live in intermediate cities or medium-sized towns. Alcosa requires cardboard boxes, bags, stationery and a number of services, which are all produced and provided by establishments in the intermediate cities. In summary, these industries play a very positive role as an employment-generation agent in the intermediate cities and the countryside, particularly important for poor households that depend on wages for a living. Bioquimex, S.A., a much larger firm, employs 150 workers in administrative tasks and 3 120 industrial workers.

A number of other organizations not working directly with the pigment firms are helping to overcome obstacles to entry into cempasúchitl production by facilitating expansion of contract farming and playing a useful role as intermediaries between firms and farmers. Asesoría y Servicios Integrados Agropecuarios (ASIA), for example, is an NGO that provides services and links farmers with input firms, banks and pigment firms. ASIA contacts BIOQUIMEX and obtains information on the firm's input requirements and willingness to buy a quantity of cempasúchitl at a certain price and date. ASIA then discusses the value of contract farming with groups of peasants; it started as a technical assistance association, so it has developed close and direct relations with them. ASIA also acts as a financial intermediary, channelling public credit to smallholders as part of the contract farming scheme, provides technical assistance and facilitates contacts with input enterprises in order to purchase inputs at a lower price.

Future prospects for the pigment industry are not auspicious. The internal market has arrived at saturation point because of the paralysing effects of the economic crisis on internal demand for poultry products. Exports could be a way out of this difficulty, but a foreign-exchange policy favouring overvaluation of the internal currency is undermining the competitiveness of national firms. Imports of Indian cempasúchitl flower are 15 percent cheaper, although of inferior quality. If overvaluation continues, firms could decide to rely more on imports, cancelling the positive employment linkages described. Technological change and productivity increases at farm and agro-industry levels, supported by a specific industrial policy, will be necessary if linkages are to be preserved and reinforced.

Poultry industry

Poultry rearing, in particular production of chicken meat, has grown by more than 500 percent in Querétaro since 1980. Because of this rapid growth, the value of poultry production in the total value of animal production in the State of Querétaro is 62 percent, according to figures from the Ministry of Agriculture. The value of chicken meat accounts for 83 percent of the total value of meat production from slaughterhouses.

This boom in poultry farming in the State of Querétaro occurred for several reasons. First, the proximity of Mexico City represents a secure market for chicken, which is in increasing demand among consumers. Second, chicken prices have until recently remained attractive, maintaining the high profitability of this activity. Third, there was inexpensive credit available during the 1989–1994 period, which allowed investment in new poultry plants and expansion of those already in existence. These plants were established as a partnership of producers in which partner participation depended on the number of chickens owned.

The economic crisis and peso devaluation that began in December 1994 nevertheless gravely affected the poultry industry. Devaluation increased production costs, because a significant part of these relate to imports, including sorghum, soya, reproductive hens and some equipment. The price of poultry increased at a lower rate and its profitability diminished. The most important factor was the heavy debt incurred by poultry farmers before the crisis began. Interest rates rose dramatically and new credit was unavailable. Poultry farmers could not carry the debt and bankruptcy was imminent.

At this point, the largest poultry company in the state, the transnational Pilgrim's Pride, bought up all the assets of the Poultry Farmers Union and absorbed their debts. As part of the pact, the poultry farmers signed contracts for six years with Pilgrim's. According to these, they committed themselves to selling their entire meat production of seven-week-old chickens to the company in exchange for balanced feed, day-old chicks, medication and technical assistance. This has converted Pilgrim's into a near monopoly. It buys between 85 percent and 90 percent of the chicken meat produced in Querétaro and exercises substantial control over poultry farmers, small and large. At present, the company sells 2 000 tons of chicken meat per week and employs 150 administrative workers and 3 120 industrial operatives. The majority of these live in the city of Querétaro, where the impact on job creation is substantial.

For poultry farmers, the contracts have represented a step backward, particularly in their independence and capacity for future growth. From being prosperous businessmen, they have been transformed into contracted producers for a large company. The contracts nevertheless have advantages for the producers, including a secure market, technical assistance, an adequate supply of genetic material and balanced food and medicine. The primary disadvantage, besides loss of business independence, is diminishing profits. This is because the price paid per kg of chicken meat is closely associated with a "meat-balanced feed" conversion factor, which has to be sufficiently high to allow for acceptable profits for Pilgrim's, which in turn demands high productivity. The problem, according to the poultry farmers, is that they are reliant on Pilgrim's for the supply of materials and they have no control over the productive process or price and quality of materials, and therefore no control over productivity and profit.

The poultry industry has important multiplying effects through employment in the region. It is the most labour-intensive animal-production activity and the one that creates the most employment opportunities, although there is a tendency to introduce capital-intensive technology. To calculate the number of jobs directly generated by the poultry industry, the farms have been divided in two types: mechanized farms accounting for 20 percent of chicken production, where one employee handles 40 000 birds, and manual or semi-mechanized farms accounting for 80 percent of chicken production, where one person attends to 12 000 birds. Based on this information and the total chicken figures in the state's poultry industry, mechanized farms generate 455 jobs and manual farms 6 045 jobs for a total of 6 500.

Poultry activity has expanded throughout the state, but there are three municipalities that constitute the largest proportion of production: Colon, with 33 percent of all poultry, El Marques with 19 percent and the municipality of Querétaro with 15 percent. The most positive effects on employment and indirectly on expense and consumption naturally occur in these municipalities, especially Colon, where there are fewer natural resources. External competition in the context of an open economy nevertheless produces pressures to mechanize production and reduce costs, to which Pilgrim's contributes greatly. This leads to the conclusion that there will be a deceleration in the creation of jobs in the regional poultry industry in future, although the jobs created are likely to require greater skills and thus provide higher wages.

Frozen and processed vegetables

Production and export of frozen vegetables is a profitable and dynamic business that creates significant employment linkages in rural towns. To understand how these linkages are created, two firms were selected and interviewed for this study. Both produce, freeze and pack vegetables for export – mainly broccoli for the American market – and for the domestic market – primarily carrots, spinach and sweet corn beans. Each uses its own technological and purveyance systems. One firm, Deshidratadora la Cascada, is a medium-sized firm established 30 years ago as a family business, which employs 160 workers and 10 technicians and produces and sells 350 000 kg of processed vegetables each year. Part of its machinery is imported, but the firm's engineers make other parts on-site. It is a very flexible firm, capable of producing and dehydrating several types of vegetables in order to take advantage of promising niche markets.

Expo-Hort, S.A., located about 25 km from Querétaro City, is a more modern firm. It was established in 1985 as an export-oriented enterprise in the context of a growing American market for frozen vegetables, particularly broccoli. Its very modern freezing machinery is completely imported and represents a substantial investment. The technology used is more capital-intensive than that used in La Cascada, but the firm nevertheless employs a significant 650 permanent workers and 50 administrative and technical employees. It produces a yearly average of US$28 million worth of vegetables, 90 percent of which is broccoli, of which US$10 million worth is exported. Its quality-control rules are strict in order to assure the product quality demanded by the American market.

Neither La Cascada nor Expo-Hort has brand names. They freeze or dehydrate vegetables, which are packed in bags with the brand name of American companies, ready for export and distribution in retail establishments in the USA. In other words, they belong to a chain that begins in Mexico's agricultural sector and ends in American supermarkets. A smaller part of production is sold to big companies such as Gerber, Campbell's, Herdez and other large firms operating in Mexico.

Both firms use completely different vegetable provision systems: direct production and contract agriculture. Expo-Hort has chosen vertical integration and produces most of its vegetables directly on privately owned farms or on rented ranches in the states of Querétaro and Guanajuato. The rented ranches are large private properties and not *ejidos*. Expo-Hort signs multi-year rental contracts with private owners. The firm invest in ferti-irrigation systems on future-rent accounts. When the land is returned to owners, they receive the equipment as part of the rental agreement. The firm selects large private properties to rent, taking into account the availability of water and the existence of good transport networks. In this way, it controls 1 600 ha of land. Only a minor proportion of its vegetable needs are obtained through contract farming.

La Cascada does not own or rent any land. It signs contracts with medium and large private farmers specifying volumes, quality specifications, prices in dollars and delivery dates. In the past, it used to work with smallholders belonging to *ejidos* but, according to the management, *ejidatarios* have problems controlling and supplying good-quality vegetables. The firm has gradually improved its list of purveyors of raw materials, who are normally modern farmers capable of controlling the quality of the product and obeying the firm's stipulation not to use certain insecticides and pesticides prohibited in the USA. They are selected within a range of 100 km of the factory to assure freshness of the vegetables. The managers said they have no supply problems.

Given the two different systems for obtaining inputs, there may be differences regarding employment generation arising from the activities of the two firms. The data collected, however, indicate that there are no substantial differences. Local demand for labour depends on the agricultural technology used, whether by the vertically integrated firm or contracted farmers. In both cases, the technology is labour-intensive. The total amount of labour involved in the cultivation of one hectare of broccoli and other vegetables is between 60 and 80 workdays per cycle in privately owned ranches, hired farms or contract agriculture, which uses more or less the same technological package.

The primary linkage effect of the export-oriented vegetables agro-industry is its demand for rural labour. The local impact of this demand is substantial: each year Expo-Hort creates employment equivalent to 128 000 workdays. The managers calculate that about 3 000 people work temporarily in the fields to produce the raw material used by the firm. They are not permanent but temporary workers, who may be employed for one month or more during the year. The workers come from Chichimequillas and other rural towns in the vicinity, where the direct impact on local income is high and expenditure linkages to local consumption shops substantial. Local transport to provide for commuting between the firm and rural towns also generates employment. In 2001, La Cascada demanded an equivalent of 32 000 workdays, a significant figure, but its impact on the local economy was less because it is spread over a larger territory. Farmers working for La Cascada through contract farming are more scattered and linkages effects are more spatially diffused. The company lacks the critical mass to generate new activities, as is the case in Chichimequillas. The impact on Querétaro City is considerably less. The company is a source of demand for specialized technical and administrative employees, who come from the area, specialized repair services, tools and cardboard boxes.

The case of large agro-industries producing vegetables for the domestic market is interesting and complements the observations made so far. A visit was paid to Gerber, a subsidiary of the multinational firm of the same name. Established in 1958 in Mexico, Gerber has a monopoly on processed infant food, with annual sales worth US$80 million. Growing demand for baby food justified establishment of a new plant in Querétaro in 1967. This intermediate city was selected because of its proximity to Mexico City, the main source of demand, and the regions where fruit and vegetables are grown.

Gerber employs 600 people in its Querétaro plant, including administrative and industrial workers, and its impact on employment is considerable because almost all of its employees live in Querétaro. It is a source of employment and expenditure linkages – salaries are relatively high – but its impact on local input industries is low, because only cardboard boxes and glass jars are purchased there. Its contribution to rural employment is not negligible, because it creates a demand for 17 000 mt of fruit and vegetables each year from Querétaro and other regions in Mexico. Spatial impact is low because its providers are scattered across various regions. Gerber prefers to deal with large-scale farmers, so direct rural impact concentrates only on 80 large producers in all of Mexico. Gerber's purveyance system is worth mentioning since only two technical employees manage the purchase of inputs from several regions for this large and complex enterprise. The secret seems to be the use of a limited list of very reliable large producers, attractive purchase prices and a sound monitoring system of agricultural production practices such as implementation of agreed rules regarding pesticide use and produce quality. "Working for Gerber has to be a good business for farmers" is the company's motto.

With regard to establishing and maintaining linkages, vegetable-export firms note the following obstacles:

- inadequate technical services from chemical laboratories checking vegetable samples;

- expensive or poor-quality inputs provided by domestic firms;

- high transaction costs resulting from excessive administrative procedures affecting imports of machinery and inputs;

- old and inappropriate export legislation.

On the other hand, Gerber does not find obstacles in the flows of agricultural raw materials; agriculture practices are linkage-friendly. The main policy recommendation from the firms' managers was to use fiscal policy to encourage new investments in labour-intensive enterprises in rural areas and towns.

Barley and beer

The beer industry in Mexico has been lucrative and dynamic for several decades and has particularly benefited from its recent successful entry into the international market. Expansion has made it necessary to enlarge the areas of cultivation of raw materials, particularly barley. To this end, the beer industry created an affiliated company, Impulsora Agricola, S.A., that is in charge of promoting and facilitating cultivation of barley and establishment of purchase agreements with producers to secure adequate supplies. Impulsora Agricola is a service company that operates in the Querétaro region and in other regions in the country, supplying the two large brewery groups, Cerveceria Modelo and the Cuauhtemoc-Moctezuma group.

To produce the seed for barley production, Impulsora has established an association with reliable farmers, whom they contract to cultivate barley under strict technical norms, supervised by the company and the Ministry of Agriculture's National Seed Inspection Service. The seeds are sold on credit, payable at the end of the harvest, to the producers who sign the contract. Farmers who have already worked satisfactorily with Impulsora in the past receive credit at preferential rates of interest. The company has a technology-transfer department to offer technical consultations to the producers free of charge. This service is offered only to groups, with the aim of reducing costs. The contracts specify the purchase of barley at prevailing international prices plus transport charges for importing barley to the plant.

Impulsora works with 300 farmers in the region of Querétaro on an extension of 4 500 ha of irrigated land. The majority are *ejidatarios* with an average of 4.5 ha per head. The effects on employment in the region are low, because barley is an almost totally mechanized crop and the breweries are situated outside the region. The positive regional effects of the brewery companies are those of offering services and a market to make the barley crop profitable. These services are important in the Mexican context because of the withdrawal of the public institutions in charge of financing and technical assistance and because of the scarcity of these services. The prices paid to the producers are nevertheless not always advantageous, because the profitability of the barley crop is equivalent or inferior to other crops in the region. In this case, the higher profitability of the large breweries does not filter down to the region, and the barley producers obtain very little of the profit generated in this agro-industrial system. This means that the indirect effects on the region through increases in agricultural income and demand for other goods are also very low.

Milk production

Milk production is the second most important activity in the region as measured by value-added and generation of employment. In order to identify the linkages generated by the agro-industrial milk system, a typology of dairy-cow producers has been established, based on information obtained from the Regional Cattle Union of Querétaro. The union is a group of 650 producers owning 35 000 cows. The milk producers have been classified as follows.

1. **Technically developed producer using full technology.** Cattle are entirely stabled (feed lots); production is mechanized from milking through to preparation of feed mixes for animals. Each farmer has an average of 450 cows; each cow gives 23 litres of milk. There are 45 large cattle farmers who own over half the total cattle stock.

2. **Semi-technically developed producer.** Cattle are stabled and farmers use mechanical milking, but food mixing and feeding of the animals are done manually. The average number of cows per producer is 100 head; their productivity is 17 litres daily per cow. There are approximately 125 producers controlling about one third of the total cattle stock.

3. **Family cattle farmers.** Producers combine cattle rearing, agriculture and other activities. Their technical level is very low and they milk by hand. The milk is for their own consumption, sale and production of homemade cheese. They do not have cooling tanks. The average number of cows per producer is ten. The cattle are reared in the farmyard or use communal lands. There are around 500 small cattle farmers representing 15 percent of the total cattle stock.

Dairy producers are linked to other local companies through the purchase of raw materials for feed, including cotton seed, soya paste or industrial waste that comes from food companies based around the city of Querétaro and other intermediate cities. These purchases are relatively low, because large and medium farmers have irrigated lands where they cultivate alfalfa, the main fodder used in the region. Family cattle farmers are almost self-sufficient in fodder and their demand for raw materials is minimal. The purchase of equipment has no regional impact, because the greater part is imported or produced in large cities.

Total employment creation due to milk production can be calculated using cows attended per full-time worker – 40 in technically developed ranches and 30 in semi-technically developed ranches. The former need approximately 500 fixed workers per year; the latter need 415, giving a total of some 915 fixed jobs in the region, a figure much lower than the 6 500 jobs generated by the poultry industry. The fact that some 670 producers and their families are partly supported by farmyard-based milk farming should nevertheless be taken into account.

The sale of fluid milk develops few linkages in the region, because more than 70 percent is sold to the ALPURA company, which is situated outside the region. ALPURA, a company to which the producers belong in their capacity as partners, picks up the milk in their cooling-tank vans and transports it to another state, where it is pasteurized and processed. Another part of the milk is sold to milk-product agro-industries. There are three companies of this type in the region producing cheese and other milk products.

The milk industry in Querétaro and all of Mexico is going through a difficult period as a result of adverse government policies. These policies have limited the expansion of the dairy industry in spite of the fact that the demand for milk and milk products keeps growing. The problem is that the Government sets a maximum price for selling milk to the public, which is fixed at a low level in order to benefit low-income consumers. With the same goal in mind, the government fixes very low tariffs on the importation of milk powder, which is very cheap on the international market. These two measures have reduced the profitability of the milk industry.

This crisis in the milk industry is closely related to the low profitability of the activity. Only producers that are integrated, mechanized and of a certain minimum scale have managed to survive well. Many small and medium producers, particularly the semi-technically equipped, have dropped out of the market. Liberalization and government policies oblige producers to resort to mechanization. Employment generation has decreased as a result of the crisis and will continue to shrink even further in the future as a consequence of capitalization. This means that there is a tradeoff between maintaining low milk prices to benefit urban consumers and maintaining and increasing rural jobs in the milk-producing regions.

Water shortages are an additional problem in Querétaro. Alfalfa, the main fodder, is a water-intensive crop; reduction of the levels of water in the region have caused the production of alfalfa to be reduced by half, contributing to a reduction in the production of milk in the region.

It is worth mentioning that family cattle farmers have survived the crisis better, because milk production is not their only activity and because they operate in a local market where they sell their products of unprocessed milk and home-made cheeses, which are produced at low input prices resulting from self-production of fodder and family labour.

Vertical integration and the linkages between companies are very important for the survival of rural producers, particularly in such a competitive and difficult market. Thanks to the existence of ALPURA, for example, Querétaro's milk producers can sell their production in the final markets at prices that guarantee their permanence in the market and without investing in the pasteurization and homogenization phases, which have heavy capital requirements. Similarly, ALPURA is a company with capacity to sell in the retail markets in several large cities across the country thanks to its extensive distribution network, which benefits the producers from Querétaro in their role as partners.

When questioned on possible policies for improving the milk industry in the region, the directors of Querétaro's Regional Cattle Farmers Union indicated the need to have more favourable macroeconomic and sectorial policies, to liberalize the price of milk and avoid flooding the domestic market with imported milk powder through a tariff policy. They indicated the need to make irrigation more efficient and to apply fertilizing-irrigation techniques; to finance these, they suggested schemes of investment with fiscal support and stimulus.

Fostering linkages: summary of the case studies

Given the potential benefits to the creation and development of linkages between farmers and the non-farm sector, this section is concluded with a review of the information presented in the case studies in order to determine how productive linkages have been formed. Table 7 examines each of the relevant markets, noting what changes have occurred within the market and the 'change agents' that have played a role in these changes.

The first conclusion that can be drawn is that in almost all cases, the agents forging links have been the agro-industrial companies, quite often in conjunction with public agencies and NGOs. In other words, the changes have not originated in the agricultural subsector, with rare exceptions, but have come from outside, generated by distributors, pigment producing companies, exporters of frozen vegetables, breweries and large poultry and milk firms. This does not mean that the farmers play a passive role. They profit from these opportunities by adopting new agricultural practices and technological innovations and by reorganizing their productive activities. Innovative farmers, who lead the way in altering practices and establishing links, play a fundamental role in these changes; their actions produce positive externalities for other farmers. The fact that the producers take rather than forge opportunities is still a potential weakness, however. This is closely related to the fact that the organizations of rural producers in the region are few and weak.

An extremely important mechanism through which the linkages are established is contract agriculture, with its various modalities. Through contracts, purchase/sale commitments are established and prices, quality and quantity requirements, input provision and other specifications are fixed. It should be

TABLE 7
Summary of linkage effects

Product	Type of new links or changes in the productive system	Change agents
Maize	Introduction of new technological package. New forms of credit and commercialization.	Ministry of Agriculture (technical assistance and introduction of fertilizers). MASECA and FIRA. Corn Club. Uniones de Ejidos.
Cempasúchitl	Expansion of production and crop change (from maize to cempasúchitl). Creation of local jobs.	Pigment companies (purchase contracts). ASIA (purchase with credit contracts and technical assistance).
Poultry	Expansion of production. Reorganization of poultry rearing due to economic crisis.	Pilgrim's (contracts that include raw materials and technical assistance).
Frozen and processed vegetables	Exportation of frozen vegetables. Technological improvement. Employment generation.	Expo-Hort (direct control). La Cascada (contract agriculture). Gerber.
Barley-beer	Expansion of exports.	Breweries through Impulsura (contract agriculture with technical assistance, seeds and credit).
Milk	Reorganization of the milk industry in the economic crisis. Capitilization and the increase in the size of companies.	ALPURA/Regional Cattle Union (vertical integration and participation, as partners in the large company).

emphasized that there are increasingly new forms of agreements, including triangular contracts in which besides the purchasing companies and producers, there are other organizations that facilitate mediation and provide services and public and private financial institutions that bring credit and guarantees. The participation of producers as partners in industrializing and commercializing companies of final products is another interesting mechanism.

POLICIES, INSTITUTIONS AND LINKAGES: EVALUATION AND RECOMMENDATIONS

The previous section highlighted the importance of the public sector and other organizations in forging new farm/non-farm linkages and strengthening existing ones. In this section, the focus shifts to examining policies that might facilitate the quantity and quality of productive farm/non-farm linkages. The section begins by discussing the institutional vacuum that has been created by the reforms of the last decade and the effects of this vacuum on farm/non-farm interaction. Other institutions and organizations are then discussed that may be useful in developing linkages. Finally, other policies that are important to promoting linkages are noted.

Institutional vacuum

Politics and public institutions play an important role in the formation of productive linkages and the generation of employment. In Mexico, the change in the model of public intervention in the economy associated with programmes of structural adjustment brought with it substantial modifications in macroeconomic and sectorial policies, in particular in public agricultural institutions. The paternalistic and interventionist policy that had prevailed for several decades was corrected. Guaranteed prices were eliminated, subsidies were substantially reduced, official technical assistance fell into disuse and many public agricultural companies disappeared, while others shrank substantially. These adjustments in economic policy-making left a large institutional vacuum that has not been filled by new public institutions. This vacuum has created an opportunity for other economic agents and organizations to take the initiative and to fill the gap little by little with new economic relations.

One illustration of this situation is the technical assistance that Ministry of Agriculture agronomists gave free to rural producers, which played an important part in the diffusion of new technologies in the region during the 1980s. With the change of policy, this service almost disappeared: only 12 percent of producers interviewed received official technical assistance in 1992, and only 1 percent received assistance from private agronomists. The lack of technical assistance occurred during the six years between 1990 and 1995, and was replaced by agrochemical and fertilizer companies or by agro-industries purchasing raw goods, such as the breweries and poultry companies.

Finally, the Ministry of Agriculture took a step back and initiated the Programa de Asistencia Técnica para Apoyar la Producción de Granos Básicos (PEAT) programme, which consisted of contracting private agronomy firms and companies to replace the official agronomists who had been dismissed. PEAT was very limited, however, because it consisted of offering assistance unconnected with other support such as financing, which impeded the producers from buying the recommended innovations. In 1997, 40 percent of producers had received technical assistance, particularly from private agronomists, a percentage that was still low. Based on survey results, 32 percent of producers claimed that the service had improved in 1997 with respect to 1992, 47 percent said that it was the same and 8 percent stated that it had worsened. Integrating technical assistance such as fertilization, fighting infestations and pests, soil conservation and water handling is an essential activity, which the government has to continue to encourage by perfecting schemes that use private agronomists. Evaluation of these schemes is important; it would be easier for the producers themselves to do this through market mechanisms. With this in mind, it is recommend here that schemes be established in which producers do not receive assistance free of charge, but make a partial payment so that they are prompted to demand efficient service adequate to their needs.

Complementing this, agrochemical shops could be transformed into purchase and technical service centres. A few of these already work in this way, offering high-quality raw materials and technical assistance to producers. With some reforms, these businesses would be ideal places for the diffusion of technology, consultation and training. Through a state-support programme, the technical capacity of these stores could be improved to offer quality services. Agreements could be reached with seed producers, commercial houses and the Ministry of Agriculture to this end. The tasks of the ministry would be to establish agreements, channel support and apply norms and vigilance.

Public agricultural credit fell considerably as a result of the reform process. At present, only one quarter of the producers surveyed are working with credit. Credit destined for purchase of machinery, equipment and installations is practically non-existent. The greater part of financing comes from the state-owned bank, BANRURAL. Only 7 percent of the producers are of the opinion that credit service has improved, however; the rest noted that credit services have worsened or remained the same. Agricultural insurance is an almost non-existent service; only 13 percent of producers interviewed are covered by an agricultural insurance policy. At present, the private banks do not represent an alternative to the public financing system.

Non-farm income earned by households represents one possible source of financing. Some saving schemes have grown up, offering small loan services to satisfy consumer needs and to support some productive activities. The only interesting development in this area has been the creation of the Credit for Administration Programme (PROCREA), put forward by FIRA. The idea is to link credit, technical assistance and commercialization, utilizing the services of integrated firms that have experience of interacting with groups of producers. The cempasúchitl programme, ASIA, is an example. ASIA provides technical assistance and functions as a mediator in purchase/sale contracts. It organizes suppliers and establishes an agreement for flower cultivation and sales to a purchasing company; FIRA opens a line of credit to finance the sowing of 2 000 ha of the flower; ASIA commits itself to giving technical assistance to the producers and offers advances for sowing and cultivation through private banks. At the end of the harvest, ASIA hands this over to the purchasing company and charges the prices stipulated in the contract, discounts the credit, pays the producers and returns the total credit to FIRA after charging the stipulated commission. It is an interesting and successful scheme that should be extended and perfected.

The withdrawal of the state-owned company CONASUPO, the primary grain purchaser in the region, has made room for the return of traditional private grain purchasers who operated in the region before CONASUPO. There is some potential for problems, however: CONASUPO was created in part because traditional purchasers, in some instances, committed abuses against small producers. The large flour company MASECA recently began to establish agreements for purchasing maize with small and medium producers, using a triangular mechanism described before as the Corn Club, in which FIRA also participated. This arrangement clearly benefits the private company and the producers have the advantage of an assured market. None of these forms of commercialization allow producers to negotiate good prices for their products, however. Their weakness in the regional grain market comes from the lack of organization and business experience among producers. Programmes such as this one supported by FIRA might include business training for producers and assistance with organization and management for producer groups.

Institutions and organizations

There are other institutions and organizations that help facilitate the exchange of resources and through which resources are exchanged. They establish connections and new economic relations, contributing to more efficient use of resources. One of these is the land market, which allows dynamic producers and those willing to invest time and resources in the agricultural activity (our stratum

0), to have access to natural capital and make use of land that could lie fallow or be inefficiently exploited. Government programmes that provide clear land titles, such as PROCEDE in the *ejido* sector, reinforce and help to develop this market; although they are not the primary force behind land transactions, they contribute to land markets by providing more transparency and confidence in agreements. Another important institution that performs a similar function is that of **sharecropping**. Through share contracts, those who decide to migrate temporarily or dedicate themselves to other activities can obtain income from their land and allow others to use it productively.

Contract agriculture is an extremely important and widely used mechanism for connecting rural producers to firms; it has the flexibility to continue this function in the future. There have been some problems of contracts not being honoured, however, particularly by the companies. To penalize these harmful practices, which create distrust in the institution among farmers, there must be changes in the legal system to make litigation quicker and less costly. Producers need more access to information on their rights, to consultations on their means of recourse and to professional services that help them defend their interests. Finally, forming producer groups of contract farming participants can help shift bargaining power from firms to producers.

Machinery rental represents another local farm-linked market of significant importance to producers with low incomes, because through this they have access to the tractors, reaping machines and heavy vehicles. Without this practice, based on confidence and local social networks, many producers could not utilize these capital goods, because they do not have the financial resources to purchase them. Actions that can facilitate development of this market would assist owners of machinery and renters.

Integrating organizations are important, because they create new links between producers, businesses and public institutions. These organizations have the potential to continue to grow and function as intermediaries by connecting services such as financing, technical assistance and commercial consultations. It is up to the state to promote these, facilitating their functions and utilizing the services that they administer most efficiently and at lowest cost. It would be worthwhile to take advantage of their comparative advantages and include them in rural-development programmes.

The importance of **producers' organizations** lies in the fact that they facilitate economic activity through establishing linkages, improving the efficiency of resource use and providing for economies of scale. In the study region, a number of *ejido* unions have worked to distribute fertilizers and negotiate agricultural insurance schemes for members. Another role of producer organizations is in assisting members to negotiate contracts with firms and facilitate beneficial contractual relationships. While organizations have the potential to play this role, they have tended to be weak and underdeveloped. Strengthening these organizations is ultimately the task of the producers themselves, but the process could be facilitated by the public sector, particularly if organizations are given an important part to play in rural development programmes.

Finally, **public organizations and rural institutions** play a fundamental role in fostering farm/non-farm linkages, because they contribute strongly to the rules of the game that dictate the interactions among agents. It is worrying that the state at present appears obsolete and inadequate: many of its structures have been inherited from a superseded model of public intervention. Institutional reform is therefore a central requirement of public policy towards the rural sector, a theme that cannot be dealt with in sufficient detail here.

Other policies

A number of activities that link the farm and non-farm sectors are particularly important to the rural economy, as are the firms that use labour-intensive technology and have strong multiplicative effects through their linkages. These firms should be considered strategic; specific programmes should be established to promote their activities, assist their capitalization and, if appropriate, the export of their products. The development and competitiveness of these firms is frequently made difficult because of incomplete sources of services, which raises their transaction costs. This is a fundamental area for design and implementation of policy measures.

Insufficient infrastructure is another potential stumbling block for rural development. In Querétaro, for example, there are not enough roads of sufficient quality for trailers to access producing areas at all times of the year, which raises the cost of transport. There is a shortage of warehouses with ventilators, and crops are frequently stored outside, which results in substantial losses. There is a scarcity of weighing machines, which increases transaction costs. Construction of roads and establishment of public services are an essential aspect of public investment that will foster development and improve linkages.

Finally, public investment in education and health is fundamental, because it reinforces human capital and raises the quality of life in the country. These services become less and less adequate in proportion to their distance from medium cities and proximity to towns and rural localities. The low-income inhabitants of marginal zones, where educational and medical services are insufficient, have to invest their meagre resources to have access to these services.

REFERENCES

COEPO. 1995. *Estudio socio-económico y demográfico del subsistema de ciudades.* Querétaro, Mexico.

CONAPO-COEPO. 1985. *Querétaro demográfico: breviario, 1985.* Querétaro, Mexico.

CONAPO. 1993. *Indicadores socio-económicos e indice de marginación municipal, 1990.* Querétaro, Mexico.

INEGI. 1996. *Sistema de cuentas nacionales de México: producto interno bruto por entidad federativa, 1993.* Mexico City.

Martner, C. 1991. *Corredores económicos regionales y transport: el caso del corredor San Juan del Rio- Querétaro.* Quéretaro, Mexico, Mexican Transport Institute. (Technical Publication No.28.)

Chapter 4
Promoting strategic linkages between the farm/non-farm sectors: the Peruvian case

Javier Escobal and Victor Agreda

INTRODUCTION

The central objective of this chapter is to analyse the linkages between commercial agriculture and agro-industry in two of the most important agricultural valleys of Peru. Recent institutional innovations in these regions have altered the relationship between potato, cotton and asparagus farmers and agribusiness. The innovations have centred principally around management experience, a critical factor in producing for agribusiness. The developments have clearly benefited participating producers, but the direct and indirect effects on employment and consumption patterns in the area are less straightforward; local upstream linkages are weak and indirect effects have largely leaked out of the region. Innovation has tended to accentuate income concentration and capital accumulation among those who are already wealthy. The exception – small-scale cotton producers – illustrates the possibility of smaller farmers purchasing management experience. "Unfriendly" linkages between the agricultural sector and agro-industry have otherwise forced small agricultural producers to develop a set of off-farm activities that are not linked with the agricultural activity as a strategy to complement their incomes.

DESCRIPTION OF THE STUDY AREAS

In 1997, a private research centre in Peru – GRADE – undertook two surveys of 30 households each in the valleys of Chincha and Mantaro. Group interviews based on a structured questionnaire were carried out to collect supplementary contextual data from five groups of 15–19 farmers each. These primary data were supplemented with information from the 1994 Peruvian Agricultural Census and the 1997 Living Standards Measurement Survey, and from interviews with inform-ants in the zone. The data collected comprised qualitative information on spatial preferences for input acquisition, contractual arrangements and income sources, and quantitative information on outputs and inputs related to the main crops examined. The survey did not collect information on consumption expenditure.

Chincha valley

The first case study was carried out in the Chincha valley, 250 km south of Lima by the coastal highway. The Chincha coastal zone is one of the most important agricultural valleys of the Peruvian southern coast; it has been linked to the cotton export market for more than a century. The zone has abundant aquifers for irrigation and plenty of flat cropland. It is dominated by the intermediate city of Chincha, whose rapid growth in the past two decades has been based on agro-industry, fishing and non-metallic mining. Chincha is near to other intermediate cities in the sierra and serves as their link to the coast.

Approximately 8 000 farms in this valley cover about 81 500 ha, of which 31 000 ha are prime agricultural land owned mainly by large landholders. Average plot size is 6.4 ha. As can be seen in Table 1, however, more than 70 percent of the farms in the valley have less than 5 ha but control only 12 percent of the land; 24 percent have less than 0.5 ha per farm.

In the Chincha valley, two types of agricultural units predominate: large and medium-sized farms, or *fundos,* managed by entrepreneurs using modern irrigation techniques, and small-scale farmers, or *parceleros,* former cooperative associates, which constitute the majority. The cutoff farm size to distinguish between *parceleros* and *fundos* is 20 ha; *parceleros* have on average 6 ha, and the *fundos* 69 ha.

Compared to the small farmers, large farmers are more commercialized and use more irrigation. Large farmers irrigate 95 percent of their land; small farmers irrigate 70 percent. Small farmers mainly irrigate with river water using the gravity method; large farmers irrigate from wells using the drip method and thus have greater water control but greater capital investment. Large farmers use more chemicals and machinery and rent more of their land; they dedicate more of their land to asparagus – 65 percent for large farmers, 8 percent for small farmers – and less to cotton – 18 percent for large farmers and 78 percent for small farmers. The rest of the small farmers' land is mainly under maize for home consumption.

Home-based non-farm activity is undertaken by 22 percent of farm

TABLE 1

Farm and land distribution by land size (%): Chincha

	Farms	Ha
Landless	2	0
< 0.5 ha	24	1
0.5 – 4.9 ha	44	11
5.0 – 19.9 ha	24	21
20 – 49.9 ha	3	0
> 50 ha	2	68
Total	8 019	81 448

Source: 1994 Peruvian Agricultural Census.

households. The probability of participation increases with farm size, probably
because larger farmers have more cash to
meet capital entry requirements. The
activities are mainly small-scale processing of cheese and yoghurt, machinery
rental, commerce and cottage manufacturing. Employment outside the home is
undertaken by 30 percent of farm households in the Chincha zone. Here the
probability of participation decreases with
farm size. Small farmers work off-farm in
agricultural wage labour and non-farm
activity with low entry requirements in
terms of education and financial capital;
examples are construction, fishing,

TABLE 2

Share of total off-farm activities: Chincha

	%
Working on other ag. land	44
Trade	16
Teaching	4
Transport	4
Household servant	3
Construction	3
Manufacturing	3
Fishing	2
Others	21
Total	100

Source: 1994 Peruvian Agricultural Census.

commerce and transport, as shown in Table 2. Large farmers tend to operate
larger-scale and capital-intensive non-farm enterprises and engage in education-
intensive non-farm salaried employment. The seasonality of off-farm labour is
concentrated during the summer, when off-farm activities peak outside the
planting season.

Mantaro valley

The second case study was carried out in the Mantaro valley, one of the most
important agricultural areas of the Peruvian sierra. The valley comprises two
distinct agro-ecological zones. The first is located on the floor of the valley and
contains the prime agricultural land and most of the population. Most land in
this area is irrigated and used for commercial purposes by farmers and by
communities specializing in potatoes, white corn, vegetables and cultivated
pasture for livestock. The second agroclimatic zone is located in the middle and
upper part of the valley; it is characterized by steep plots, most of which are not
irrigated. Agricultural activities are commercial and for self-subsistence. This is
the most important area of the valley in terms of production of tubers and cereals –
mostly potatoes and barley; it is occupied mainly by peasant communities. The
upper part of the valley was historically owned by *haciendas* (large farm estates)
devoted to sheep-raising. These *haciendas* were expropriated in the 1970s under
the agrarian reform and the land was occupied by cooperative farms. Most of
these cooperatives have been parcelized during the last decade.

The study area was chosen to include several districts of Huancayo province where over 37 000 commercial farmers and peasants own 308 000 ha. As shown in Table 3, most farms have less than 5 ha and 42 percent have less than 0.5 ha each. Holdings are highly fragmented: fewer than 5 percent of farmers in Huancayo have only one plot, whereas 67 percent have five or more plots. Average plot size is 6 ha, but of these only 1 ha is agricultural land. One third of agricultural land is irrigated.

TABLE 3
Farm and land distribution by land size (%): Huancayo

	Farms	Ha
Landless	0	0
< 0.5 ha	42	1
0.5 – 4.9 ha	51	10
5.0 – 19.9 ha	6	6
20 – 49.9 ha	1	2
> 50 ha	1	81
Total	37 117	308 067

Source: 1994 Peruvian Agricultural Census.

The city of Huancayo, founded in 1572, is located in the centre of the corridor that connects the towns of Huancavelica and parts of Apurimas and Ayacucho. It is thus an obligatory stop for all agricultural production sent to market in Lima, and for all goods entering these departments from the capital and the rest of the country. This has led to Huancayo's position as a commercial centre, but the industrial sector – particularly the agro-industrial element – is little developed. Lack of an agro-industrial sector may be a result of proximity to Lima and its wholesale markets, or of the absence of an appropriate agricultural product. The few private and public attempts have failed or achieved limited success, such as dairy, canned and milled products. Private potato processing farms have only recently begun to take an interest in maintaining ties with local producers, a phenomenon discussed in more detail below.

Three districts in close proximity to Huancayo were chosen for study. Huayucachi is located 11 km south of Huancayo; including outlying communities, it has a population of 1 500 households. Farmers in Huayucachi produce potatoes on communal land; they have strong links with wholesalers at the La Parada market in Lima in terms of potatoes for consumption, and with potato producers on the Central Coast in terms of potato seeds. The population has grown in recent years as a result of migration as families from outlying communities sought refuge and protection from the political violence of the 1990s. This migration has furthered the urbanization of Huayucachi in terms of housing and small businesses.

Huanchac, which consists of 480 families, is also located in the floor of the Mantaro Valley, 28 km from the city of Huancayo. Because of the availability of irrigated land, agricultural production focuses on potatoes and vegetables; the

TABLE 4

Land, production and market insertion: Huancayo

	Huayucachi	Huanchac	Aramachay
Average cultivated land area (ha)	13	10	6
Land characteristics (%)			
Rented	37	30	15
Irrigated	66	77	19
Potato production	42	60	33
Vegetable production	15	23	Ns
Average potato production, by farm (mt)			
Total	117	124	17
Consumption	78	73	15
Seeds	33	7	1
Processing	6	44	1
Market destination (%)			
Consumption, sold in Lima wholesale market	86	92	16
Consumption, sold in local market	12	6	59
Consumption, kept for home consumption	2	2	25
Seed, sold to coastal producers	95	20	Ns
Processing, sold to processing industry	5	70	Ns

Ns = not significant.

Source: GRADE field work, 1996–1997.

microclimate and soils are good, and two potato-processing companies based in Lima have recently started working with producers from Huanchac. The areas of Huayucachi and Huanchac are characterized by good access to public services and infrastructure; the proximity of the city of Huancayo allows for commuting to employment.

Aramachay represents the districts located in the upper part of the Mantaro valley. Its distance from Huancayo – 64 kilometres – has resulted in sparse provision of public services and infrastructure. Agriculture, still organized along communal lines, is the most important economic activity; potatoes and barley are the principal crops; families also depend on livestock activities and seasonal agricultural wage labour. As can be seen in Table 4, producers in Aramachay have smaller plots than the other two case-study districts and much less access to irrigated land. Combined with lower yields, Aramachay farmers produce on average only a sixth of the total average production of farmers in Huayucachi and Huanchac. Most of it is sent to local markets or consumed on-farm; farmers in Huayucachi and Huanchac export almost all of their production outside the region.

Although agriculture is the core activity of Mantaro farmers, over 24 percent of them acknowledge undertaking non-agricultural activities on their farms; the number of such farmers increases as the size of agricultural assets diminishes. The main on-farm non-agricultural activities are handicrafts and trade. Off-farm activities are also important in the Huancayo Province, especially outside the September–December planting season. The most important off-farm activities according to the 1994 Agricultural Census are listed in Table 5.

TABLE 5
Share of total off-farm activities: Huancayo

	%
Working on other ag. land	23
Trade	16
Household servant	12
Transport	11
Construction	6
Teaching	5
Mining	4
Manufacturing	1
Others	21
Total	100

Source: 1994 Peruvian Agricultural Census.

The Aramachay district offers few off-farm employment opportunities; this, combined with the long distance from Huancayo, means that most households work and live on-farm. The opposite is true for the other two districts, where urban growth and the proximity of Huancayo provide multiple off-farm employment opportunities and the majority of households do not live on-farm. Even though agricultural production is the principal economic activity for 62 percent of households in Huayucachi, 75 percent also work off-farm. Poor families in Huayucachi work in a variety of activities, including unskilled wage employment, shoe making and making and selling vehicle parts. In Aramachay, 90 percent of households work in agriculture but only 25 percent work off-farm.

INSTITUTIONAL INNOVATION AND LINKAGES

Chincha valley

Cotton

Cotton was the motor of economic and town growth in the Chincha valley for most of the last century. In the past decade, however, the fortunes of cotton have been declining as a result of appreciation of the real exchange rate, competition from liberalized imports of textiles and cotton and increasing input costs for farmers resulting from cuts in government subsidies for inputs and credit. These factors provoked a national slide in cotton production, from 293 000 mt in 1989 to 86 200 mt in 1998 (Ministry of Agriculture, 1999).

As cotton became less profitable, non-traditional crops emerged as more profitable options. This profit difference and other institutional factors discussed below provoked a massive shift of large farmers out of cotton and into asparagus, oranges, apples, avocados and lucuma. This left room for small farmers to enter cotton production as a less profitable but less demanding crop in terms of organization and capital requirements.

There have been some recent instances of vertical integration of ginning and textile manufacturing firms, but the usual organization has been separate firms. The ginners act as intermediaries, buying raw cotton from farmers without agreeing contracts, and process the cotton into fibre and oil seeds to sell to textile firms and edible-oil factories. When large farmers produced cotton, they either sold to ginners acting as intermediaries for the textile firms or bought ginning services and sold directly to the textile firms.

When small farmers moved into cotton, they obtained credit from ginning firms or from large cotton growers and sold the raw cotton to them. When large farms shifted away from cotton, the small farmers became part of a system in which the ginners were almost the only source of credit, which was provided in-kind in seeds and chemical inputs.

From about 1995, difficulties increased for small farmers participating in the cotton subsector. Structural adjustments reduced access and increased costs for inputs and credit. The cost increases were magnified by the disappearance of the cooperatives, which had been dismantled in the 1980s. Ginners stepped in to fill the input credit gap left by government withdrawal, but at rates well above the former state-subsidized rates.

To offset rising input costs, small farmers turned to NGOs and rural and municipal savings/credit schemes. NGOs offer technical production assistance and credit at below market cost, subsidized mainly by foreign donors. These new sources of credit reduced small farmers' dependence on the ginners' expensive credit. The NGOs also negotiated with the ginning firms, with varied success, to increase the price for raw cotton.

The coverage of NGO schemes was partial and left out numerous smallholders. The schemes were limited to reducing credit costs but did not address numerous other problems faced by small cotton farmers, in particular lack of marketing and negotiation expertise to deal with other cotton-chain actors, expensive variable inputs and lack of organizational capital that was formerly embodied in cooperatives, which permitted economies of scale in input and credit acquisition.

These gaps in human and organizational capital and high transaction costs created a constraint and the opportunity for innovation to meet a need. The latter arose in the form of endogenous private institutional innovation, a share-tenancy arrangement remarkably similar to that described theoretically by Eswaran and Kotwal (1985). In 1999, a local large farmer established what can be called a management company. The company sells management services to small cotton farmers in return for a 25 percent share of the profits from cotton sales.

The management company requires the formation of farmer companies. The management company negotiates a contract with each farmer company that involves production and marketing actions such as timing of input use and bulk purchase of inputs, and group acquisition of bank credit with the manager's intermediation and the farmers' land as collateral. By the end of 1999, ten farmer companies were in operation, involving over 400 farmers.

This arrangement led to reduction of transaction costs and economies of scale in input purchase and product marketing, and to formal input market transactions. The latter allows a farmer company to abandon tax exoneration. When a farmer company uses the system of tax exoneration, it can no longer use the tax it pays on the purchase of inputs as fiscal credit. At the same time, the textile firm punishes the exonerated firm, because receipts can not be discounted according to the Peruvian value-added system when it buys its cotton.

The arrangement allows purchase of inputs in bulk, transported in trucks rented by the management company, which overcomes a physical capital constraint. This allows scale economies in input acquisition and purchase of inputs in Lima at significant savings compared to buying from local input dealers. Beyond these services, the manager is planning to help the client farmers with diversification of their product mix to reduce risk.

The arrangement has encouraged changes in the organization of the cotton subsector in Chincha. The farmers' companies now contract ginning services and sell the ginned cotton directly to the textile firms. The companies get better prices than before their contract with the management company, which increases profit. Other cotton farmers in Chincha, numbering nearly 1 000, sell their raw cotton to ginners at a disadvantage compared to the farmers' companies.

The present survey revealed that profitable options for small farmers have narrowed to working with an NGO or with the management company; farmers outside these arrangements are going out of business and selling their land, or are taking one of these two survival options. This appears to be mainly because even after paying the substantial management fee, profit when working with the management company is still greater than farm profits in NGO-organized schemes. The profit/cost rate is 80 percent higher in this tenancy contract as compared to working alone and about 50 percent higher than the NGO option as a result of the use of the tax credit. The NGO arrangement is being fuelled by donor funds, so it is less endogenous and probably less sustainable in the long term.

Asparagus

Many large farmers in the valley moving out of growing cotton have in recent years switched to asparagus, fuelling a production boom. This boom emerged for several reasons. Peru's southern coast provides an exceptionally good climate for asparagus, offering producers two harvests per year. The resulting yields of 12 000 stalks/ha are far superior to the 7 000 stalks/ha registered in Spain, Peru's main competitor and one of the foremost producers and consumers of asparagus. Labour costs in Chincha are low compared to those of Spain; irrigated farmland with registered titles is abundant. Chincha is close to the Lima market. In the past decade, there has been a virtual elimination of political violence in the region. Finally, the heritage of the long cotton boom is a large number of experienced farmers, agronomists and input and transport firms who are geared to commercial agriculture.

The relationship between asparagus agribusiness firms and the *fundos* is formalized in a contract. The company agrees to provide credit, technical assistance and inputs, and operates fixed pricing rules. The producers in turn promise to reserve all output for the company, following strict production guidelines, and to allow supervision by company technicians. The firms also buy asparagus from small farmers, but the latter do not enjoy the benefits of contracts and the price they receive is usually much lower as a result of quality differences. The great majority of small farmers cannot meet the requirements of quality asparagus production, because they lack managerial and technical expertise. The asparagus agro-industrial firms have a strong preference for contracting with large farmers because of monitoring costs and the capacity constraints of small farmers.

Interviews with the asparagus companies nevertheless revealed that they would like to increase production of asparagus on their own lands, obviously outside the contract system. This has recently been made possible by yet another institutional change: the law was recently overturned that had made it illegal for agro-industrial firms to own cropland and thus vertically integrate. This may subsequently undermine contract agriculture in one of the few places in Peru where it appears to have been functioning well (see Figueroa, 1996).

The advent of the asparagus agro-industry required and brought two institutional changes, the first leading to the second. The first change was in the institutions of grades and standards. The canned/bottled asparagus export market from Peru is highly demanding in terms of quality and safety standards with certification schemes by the Peruvian export association (Diaz, 1999).

The second institutional change was driven by the stringent standards and technological and capital demands involved in asparagus production. The advent of participation in asparagus export brought with it the institutional change of requirements for stringent quality and safety standards, which in turn induced further institutional change in terms of emergence of agro-industry/farm contracts to assure compliance with the standards. These contracts had not previously existed in the Chincha valley because cotton, the dominant crop of the large farmers in the past, had not been produced and sold by contract.

Direct effects

The direct employment impact of agro-industrialization includes employment in participating farms resulting from changes in product composition, technology and scale of production, and employment in agro-industrial firms. These effects are conditioned by the extent to which agro-industrial firms outsource their intermediate inputs or produce their own, for example whether textile firms use imported cotton or buy local cotton, and by the technology and scale differences implied by these alternatives. The effects of cotton and asparagus agro-industrialization in Chincha and institutional change flowing from it are hypothesized as follows.

First, the shift in the past decade by large farmers from cotton to asparagus has among other things tended to reduce the demand for farm wage employment and for direct farm labour per unit of agro-industrial output. Asparagus production is only 25 percent as labour-intensive as cotton production; asparagus agro-industry favours links with large farmers rather than small farmers who tend to have higher labour/land ratios; the labour/output ratio in the cotton agro-industry is about twice that of the asparagus agro-industry.

Second, there is a counterbalancing increase in employment from the increase in small-farmer cotton production, because cotton is more labour-intensive than subsistence maize. The increase in profitability from the institutional innovation discussed above would magnify and sustain this increase.

Third, the two effects on employment act in opposite directions: asparagus agro-industrialization and institutional change imply a drop in local employment, whereas there is an increase in the case of cotton. A rough calculation suggests nonetheless that net aggregate effects are probably negative regarding incomes of small farmers, with a drop of about 8–9 percent; the potential drop was substantially buffered, however, by the endogenous institutional innovation in cotton. The reasoning for this is shown in the following two steps.

Based on survey information showing that the change from cotton to asparagus involved roughly 1 700 ha and on corresponding labour/output ratios, the change produced a 6.6 percent drop in agricultural employment, and a second-round effect in the industry of a 25 percent drop in employment at the agro-industry level. Overall, these figures meant an 18 percent drop in employment income due to the change.

On the other hand, the income increase from the increase in cotton-production profitability for small farmers resulting from the institutional innovation is roughly 10–15 percent. That income increase is based on the following information about net gains from working with the management company. The company charges 25 percent of farmer direct costs, which works out to approximately US$239/ ha. The benefits can be described as savings of costs relative to what small farmers paid before entering into this new arrangement:

- reducing input prices – 24 percent relative to prices normally paid at the shops in Chincha for reasons discussed above – saving US$116/ha;

- reducing loan interest by going to the local bank as a group rather than to the ginning firm for credit, saving about US$90/ha;

- obtaining a better price for the cotton by now selling directly to textile companies after contracting ginning services, generating gains of about US$150/ha;

- company status allows a change in tax status, allowing them to benefit fully from the value-added tax system in Peru, which allows savings of US$180/ ha.

The net effect is that the small farmers spend US$239/ha for the management service, but save US$536/ha with new system, thus netting US$297/ha with the new system.

Indirect effects

The indirect employment effects include employment from net output changes in businesses in production-linkages forward and backward from farms and agro-industrial firms, and from consumption expenditure linkages from incomes earned in farms and agro-industrial firms. Table 6 shows the use of inputs on asparagus on large farms and on cotton on small farms. While small cotton farmers are much more commercialized and technologically equipped than subsistence maize or potato farmers in Peru, there are nevertheless substantial differences in the technologies and acquisition practices of small and large farmers in Chincha; their impacts on the local economy are therefore different.

First, nearly all farms use fertilizers, but the use rate per ha is much higher on asparagus than cotton. All farms use herbicides, insecticides and fungicides. The difference for the local economy is that asparagus farmers buy chemical

TABLE 6
Input use in *fundos* (asparagus) and *parceleros* (cotton): Chincha

	Fundos	*Parceleros*
Fertilizer use (%)	100	95
Use per ha (kg)		
Ammonium sulphate	500	200
Ammonium phosphate	300	180
Potassium phosphate	400	100
Main purchase location	Lima	Chincha
Hybrid seed use (%)	100	100
Seed use per ha. (no. of stalks/kg)	20	1 800
Seed origin (%)		
Own production	25	25
Bought	75	75
Main purchase location	Chincha/Ica	Chincha
Herbicide, insecticide and/or fungicide use	100	100
Main purchase location	Lima	Chincha
Machinery repair	Chincha	Chincha
New machinery, spare parts and tools purchases	Lima	Chincha/Lima

Source: GRADE field work, 1996–1997.

inputs from input firms in Lima, whereas cotton farmers buy from a merchant in Chincha; this has started to change, however, because the farmer companies can buy in bulk from Lima. Why the difference? Large farmers have the asset base to serve as collateral and the management capacity to make contacts and to rent vehicles to go to Lima, where they get lower prices for larger lots. By comparison, farmers who have to buy from Chincha input firms are forced to buy at higher mark-ups because of lack of competition.

Second, asparagus and cotton producers produce hybrids. Comparing seed and seedling use rates is not comparing like with like; it should be noted, however, that the purchase rate is the same at 75 percent, so these small farmers are relatively technically advanced compared to subsistence farmers in other parts of Peru. Both are bought locally.

Third, both kinds of farmers use mechanics in Chincha, but for tools, machines and spare parts large farmers tend to go to stores in Lima and the small farmers to stores in Chincha or to cottage-manufacture workshops. It was found that the institutional change in the cotton subsector reinforces this tendency, because the cotton farmer companies prefer to buy in Lima. This may be changing in the medium term, however, as the recent follow-up informal survey shows that local Chincha firms are trying to compete with Lima firms and offering similar bulk deals.

A check for the difference in technology between the two kinds of farmer shows that there are roughly similar spatial acquisition patterns for inputs. The smaller/poorer farmers buy inputs locally and thus benefit the local economy through upstream production linkages, at least to local commerce. The medium/large farmers buy inputs in Lima and thus obtain better-known brands, lower unit prices, greater information from the dealer, better product quality and quality guarantees, greater diversity of product, larger lots and perhaps more up-to-date equipment. But they gain all this while bypassing the local economy.

Fourth, asparagus firms have an internalization policy and so do part of their own transport and repairs, thus reducing linkages to the local economy. This is mainly because they perceived high monitoring cost in key parts of their production process.

Fifth, although the data were not avaliable for evaluating consumption expenditure effects specifically in the households surveyed as part of the agro-industrialization study, it was possible to use data relating to the study area from the 1997 Living Standard Measurement Study (LSMS) survey. The data show that the richer the rural household, the higher the share of non-farm expenditures in the total, as expected from Engel's law. The 20 percent of poorest households have a share of non-food expenditures of 51 percent; the richest 20 percent have a share of 58 percent. The share of processed-food items in total food expenditure is, for the same area, 17 percent in the 20 percent of poorest households and 27 percent in the richest 20 percent. Most non-farm products are purchased in intermediate cities and consist of modern manufactured goods. Hence, as richer households are earning the asparagus profits, expenditure effects tend to benefit the intermediate city and Lima rather than the rural areas. There is historical evidence of such an effect: the growth of the city of Chincha was linked to the previous agro-industrial boom in cotton as large farmers benefited; development is now further increased by the boom in asparagus.

Mantaro valley

Potatoes

Three types of linkages between potato producers and the market were identified.

1. Retail and wholesale markets for potatoes for consumption. First, potatoes for consumption are the principal crop for most producers in the Mantaro valley. As can be seen in Table 4, producers located on the valley floor in Huayacuchi and Huanchac sell to wholesale markets in Lima. Producers located in the higher parts of the valley instead sell at local markets to intermediaries who channel the production to wholesale markets.

2. Coastal producers of potato seeds for sale. An estimated 14 500 mt of potatoes for seed are produced in the Sierra for farmers located in the central and south coast, supplying seed for over 8 000 ha. This part of the Sierra is known as a traditional producer of quality seed, so coastal producers and intermediaries establish long-term relationships with the seed producers, an arrangement which often includes provision of credit.

3. Potato processing companies. This link has developed relatively recently in the valley as a result of growth in demand for processed potatoes in Lima and the rest of the country. The most important company is Savoy Brands Peru SA, a transnational in which 85 percent of all sales of processed potato products are concentrated. Until 1994, supply for this company was obtained through intermediaries, among whom were the wholesalers of La Parada market. Since then a direct relationship has been established with producers.

From August to November, the company typically derives its supply of the Tomasa Condemayta variety from medium-sized producers in the coastal valleys

TABLE 7

Ties with coastal producers for the sale of potato seeds: Huayucachi

	Seed producer
Average production (metric tonnes)	
Total potato production	117
- home consumption	78
- potato seeds	33
- processed potatoes	6
Market destination (%)	
Consumption, sold in Lima wholesale market	86
Consumption, sold in local market	12
Consumption, kept for home consumption	2
Seed, sold to coastal producers	95
Processing, sold to processing industry	5

Source: GRADE field work, 1996–1997.

near Lima. This variety has the best traits for processing potato chips – low sugar and water con-tent and appropriate uniform size. Since the traditional potato varieties in the Sierra do not have these qualities, for the other months of the year the company must stock up on coastal potatoes and store them in rented refrigerated warehouses. Even so, supply is not sufficient and the company often im-ports the Diacol Capiro variety from Colombia.

In this situation, the company established a relationship in 1995 with Mantaro Valley producers for supply during critical months. The Mantaro valley was chosen because of its proximity to Lima and its good transport links. The end of political violence in the area and most importantly the managerial experience developed by Valley potato producers were contributory factors. The company selected 50 producers who control over 1 000 ha in production, with average yields above 20 mt/ha. Most of the seed producers are professional technicians with years of experience in seed production. The principal characteristics of these producers are shown in Table 7.

The company provides the seed producers with seedlings, with the objective of producing *tuberculillos prebasicos* of the Diacol Capiro variety in greenhouses. The company guarantees the purchase of a specified quantity of *tuberculillos* at a fixed price and quality. The company then passes the *tuberculillos* to specific-consumption potato producers in the valley, to whom the company guarantees the purchase of output at a fixed price. For their part, seed and consumption producers give exclusive rights to the company for their output, and allow company technicians to inspect production processes.

Direct effects

The direct effect on employment of the advent of institutional change in potato contracts is more straightforward than in the Chincha case. The new variety produced in Mantaro has tended to reduce the demand for farm wage employment. First, the new variety is less labour intensive than traditional Sierra varieties; second, potato agro-industry favours large farmers, who tend to have lower labour/land ratios. There is no counterbalancing increase in employment in the small-farm sector, because innovation is limited exclusively to the technically most advanced and experienced potato farmers. Little positive direct effect in employment is foreseen, and it may turn out to be negative.

Indirect effects

As in the Chincha case, the indirect employment effects include employment from net output changes in businesses in terms of production-linkages forward and backward from farms and agro-industrial firms, and from consumption expenditure linkages from incomes earned in farms and agro-industrial firms. Table 8 shows the use of inputs on farms in the three districts. Many differences emerge, which have important implications for the existence and strength of local linkages.

First, 90 percent of farmers in Huayucachi and Huanchac use fertilizer, a higher share than the 60 percent in Aramachay, and much more intensively. Almost all producers use herbicides, insecticides or fungicides. As with Chincha, the difference to the local economy is that farmers from Huayucachi and Huanchac tend to buy their inputs from commercial establishments in Huancayo or Lima; Aramachay farmers buy theirs from local merchants. As with Chincha, the larger Huayucachi and Huanchac farmers have the asset base to serve as collateral and the management capacity to make contacts and rent vehicles to go to Lima, where they pay lower prices by buying in larger lots. By comparison, Aramachay farmers who have to buy from local input firms are forced to buy at higher mark-ups because of lack of competition.

Second, all farms in Huayucachi and Huanchac use hybrids, mostly purchased from valley seed producers. The institutional innovation thus takes advantage of a long relationship between local potato farmers and seed producers. Most Aramachay farmers use hybrids as well, but rely much more on their own production of seeds.

TABLE 8
Local linkages in potato production: Huancayo

	Huayucachi	Huanchac	Aramachay
Fertilizer use (%)	90	90	60
Use per hectare (kg)			
Urea	310	350	155
Super triple	219	250	110
Potassium phosphate	133	124	66
Main purchase location	Huancayo	Huancayo	local/Huancayo
Hybrid seed use (%)	100	100	78
Seed origin (%)			
Own production	25	25	75
Bought	75	75	25
Main purchase location	Valley seed producers	Valley seed producers	local market
Herbicide, insecticide			
and fungicide use	100	100	85
Main purchase location	Huancayo/Lima	Huancayo/Lima	local/Huancayo
Machinery repair	Huancayo/Lima	Huanchac/Huancayo	Aramachay/Huancayo
New machinery, spares			
and tools purchases	Huancayo/Lima	Huancayo/Lima	Aramachay/Huancayo

Source: GRADE field work, 1996–1997.

Third, all farms tend to use local mechanics, but Huayucachi and Huanchac farmers rely more on Lima businesses for tools, machines and spare parts; Aramachay farmers depend almost exclusively on local purchases. Farms in Huayucachi and Huanchac prefer well known brands for tools; farms in Aramachay tend to use locally produced tools from cottage-manufacture workshops.

Thus two segmented markets are evident, with correspondingly different impact on the local economy. Aramachay farmers, who are generally smaller and poorer, buy inputs locally and thus benefit the local economy through upstream production linkages. The Huayucachi and Huanchac farmers, who are larger and wealthier, buy inputs in Lima and thus obtain more known brands, pay lower unit prices, receive more information from dealers and have better product quality and quality guarantees, greater diversity of product, larger lots and perhaps more up-to-date equipment. But they gain all this while bypassing the local economy. Institutional innovation has had little indirect impact on the local economy.

Finally, a consumption impact similar to Chincha should be expected. Only wealthier farms are benefiting from the institutional innovation in Mantaro valley; these households will spend a higher share of income on non-food expenditures, which are purchased in intermediate cities and consist of modern manufactured goods. Because richer households are earning the profits from processed potatoes, expenditure effects tend to benefit the intermediate city and Lima rather than local rural areas.

CONCLUSION AND LESSONS

Chincha

In contrast to other valleys, the significant presence of modern farming equipment in Chincha guaranteed to industrial firms the ability to implement productive technologies and ensured the availability of high-quality raw materials. Factors contributing to the success of linkages between agro-industry and local asparagus growers include sufficient agricultural knowledge and management capacity of the majority of landowners, availability of modern technical equipment and availability of arable land – on average over 50 ha per holding – most of which is equipped with irrigation systems.

These agreements are established through contracts in which the local producers obtain exceptional market conditions including rent contracts, credit and technical assistance, resulting in reports of higher earnings with respect to those in cotton production. Nonetheless, in this type of arrangement between agro-industry and local growers, the small landowner does not receive the direct benefits that are generated. In the end, they benefit solely as providers of labour during harvest seasons.

As in other areas of the country, such as San Lorenzo with its mango producers, the asparagus industry in the Chincha valley continues to buy up local land and to opt for vertical integration. This process has not yet ended, but it is possible to affirm that agro-industrialists have opted to ensure access to quality raw materials for two reasons. First, many asparagus plantations have already entered a phase of diminishing yields, and their infrastructure needs renovation; this also presents an opportunity to renew contracts with landowners under a new arrangement. And until recently, outside businesses could not buy out local landowners; under new legislation it is now possible. Second, there is a need for additional areas for cultivation beyond the land in the hands of middle-sized landowners. Given this shortage, sufficient contracted lands are no longer guaranteed to agribusinesses, which are therefore now interested in buying up land. It is clear that new legislation has offered large firms greater bargaining power in their negotiations with middle-sized landowners, but the situation has worsened from the viewpoint of local landowners. In order to assure a continuation of the beneficial contracts they receive from agribusinesses, middle-sized asparagus producers must now be more efficient in their operations, because firms in the asparagus industry now have a reasonable indicator of the cost of producing 1 kg of asparagus in their own fields to compare with the prices of local producers.

In Chincha, the abandonment of cotton production on middle-sized and large land holdings in favour of asparagus crops has consequently meant that small landowners are the principal cotton producers in the valley. A recent phenomenon in this valley has been the recent emergence of contractual arrangements between a small private consulting firm and an important group of producers, with the objective of augmenting the participatory role of local growers in the commercial aspects of cotton production. This arrangement has brought significant economic benefits to small cotton producers.

With this method of operations, the management company has identified a role that is profitable for them and that generates a circle of benefits, allowing local economic agents to interact with greater ease and thereby increase overall utility. Under this arrangement, private banks are more willing to offer loans to producers on the basis of the reputation, experience and knowledge of the owners of the consulting agency supporting them – provided, of course, that each small farmer complies with the bank's minimum requirements in terms of property titles, collateral and down payments. By connecting themselves with the matrix of commercial activity of input producers located in Lima, they can acquire significant price discounts for large-volume purchases. By gaining direct access to the textile companies, they receive higher prices with respect to those previously received from intermediaries. In other words, this arrangement generates economies of scale in production that individual farmers are unable to achieve alone, thereby providing economic gains for local growers as well as the private management consultants.

Mantaro valley

Processing potato-based snack foods is one of the most dynamic and important areas of activity in the market for processed potatoes, in which important changes are occurring as a result of private producers' efforts to improve their market position and increase sales. The most important firms involved in this line of products are multinational affiliates. Until three years ago, the system used by these firms to obtain supplies of potatoes used intermediaries; it is now based in a direct relationship between multinationals and groups of producers that varies according to seasons of production.

In the Mantaro valley, contracts require farmers to have sufficient quality land, irrigation, human capital and managerial skills to meet demanding production and marketing schedules, to sell only to the contracting firm and to submit to that firm's technical supervision. Only large farmers in the Mantaro valley can meet such requirements. The rewards, through the contracts, are technical assistance, credit for land rental and input acquisition, quality seedlings supplied by the company at agreed prices, and profitability and risk reduction at a fixed price adjusted between contracts depending on international prices.

The valley's proximity to Lima, the present environment of security and the transport facility provided by some of the country's best highways make this valley a desirable location for acquiring agricultural inputs. The presence of leading producers with significant managerial capacity is of great importance. The majority of corresponding seed growers are professional technicians with years of experience in seed production who use areas of cultivation located in agro-ecological zones that are appropriate for seed development. Similarly, the selected potato growers are producers specialized in this crop, whose production has been traditionally destined for the wholesale market of La Parada.

Lessons

It is clear that access to public goods and services coupled with the appropriate amount of private assets, especially education and managerial ability, can dramatically improve the quality of farm/off-farm linkages. In this respect, improvement of public roads in a country like Peru is a significant element in reducing marketing costs. Improvements in land titling and registration is important as a means to increase the chances of obtaining credit.

In each of the two cases analysed, however, these elements have been necessary but clearly not sufficient to improve linkages between small and medium-sized farms and agro-industry. Managerial abilities are a crucial element to developing successful linkages; this paper has focused on the phenomenon of endogenous institutional innovation in two areas of Peru involving traditional crops – cotton and potatoes – and a non-traditional crop – asparagus. These changes have been induced by changes in the general institutional context such as the emergence of demanding quality and safety standards in agro-export markets and domestic agroprocessing markets, the policy and market context in terms of the withdrawal of government support to input and credit markets, and the factor-distribution context – the scarcity of management and technical and marketing expertise among small farmers.

The institutional innovations that emerged in the zone were twofold. The first included contracts between agro-industrial firms and large farmers; the contracts were introduced by the firms to assure timely delivery and compliance with strict requirements required by the new and demanding quality and safety standards for agro-export of processed asparagus and processed potatoes. The second included management services exchanged for labour supervision and land collateral in share-tenancy contracts between a management company and farmer companies of small cotton farmers.

The importance of these institutional changes is twofold. First, they were induced institutional innovations driven by the requirements of agro-industrialization itself; second, they had ambiguous employment and income impacts that tended to be negative. In Chincha, on the one hand, the emergence of asparagus and firm/farm contracts reduced employment through exclusion of small farms and shifts to capital-intensive crops. On the other hand, the reinforcement of smallholder cotton and the emergence of farmer companies increased smallholder employment and incomes. The institutional innovation allowed them to reduce risk and increase profits, and thus access some of the benefits of agro-industrialization and globalization. Processing firm/farm contracts are common in Peru, as is the presence of NGOs bringing subsidized credit, but the private management firm innovation is rare and new in Peru, and apparently new in the region. Policymakers and NGOs have recently discovered that this innovation is taking place; they are asking hard questions about whether the innovation can and will be diffused. The interest in the private for-profit institutional change is sharpened by growing doubts about how economically sustainable and widespread NGO help can be for small farmers in maintaining their participation in income-enhancing agro-industrialization. With changes in land laws and markets, the fluidity of the situation is apparent; agro-industrial firms are even starting to ask themselves whether contracts with large farms are necessary or the best option.

Two policy issues emerge. First, can policymakers do anything to facilitate the emergence of such private institutions, which are profitable for suppliers and customers but do not jeopardize their inherent strength – their endogeneity? Second, can policymakers undertake complementary actions to facilitate agro-industrialization in demanding profitable sectors?

There may be a role for the government. It may be a direct role, or it may involve temporarily subsidizing or providing facilities for private management, accounting and technical training for farmers in zones already undergoing agro-industrialization or that have potential for it. The objective would be to improve the conditions for the emergence and low-transaction-cost functioning of such private institutions among management-service providers and the farmer companies. These policies should be aimed at reducing transaction costs for the emergence and development of managerial services that will be demanded only if they achieve a superior input mix through resource pooling in the face of a moral hazard problem.

On the other hand, the larger farmers, who have better land and access to education and water, were ready to profit from the shift from cotton to the more profitable asparagus in Chincha, and to processed potatoes in Mantaro. Among the small farmers in Chincha, those with titled land were better positioned to join the lucrative farmer companies. It is thus important that there be renewed attention to building the private and public asset base of small farmers in these areas. That is crucial to their participating in agro-industrialization rather than being excluded from it. The World Bank (1998) discusses examples of these agro-industry facilitation actions in other zones of Peru, including provision or facilitation of the private development of physical capital such as wells and nurseries, managerial capital such as technical expertise and management experience, infrastructure such as roads to growing markets and input sources and essential public services such as resolution of problems related to plant and animal health or the availability of registered land titles.

Finally, at least basic preliminary evidence has been presented that the direct employment effects of agro-industrialization in the Chincha valley are ambiguous and tend toward income concentration. The same process is probable in the Mantaro valley. It seems clear, however, that the indirect spin-off effects of the process point in the direction of concentration of gains among farmers who are already better off: the larger farm households that dominate the off-farm manufacturing and service sectors in rural areas and local intermediate centres. Upstream production linkages are correspondingly weak.

There is a need to recognize that these impacts have been concentrated and to seek a broadening of participation by poorer groups. This would result in more equitable development, and smaller producers would tend to have stronger local production and consumption linkages. Training, technical assistance and credit provision for small and medium-sized enterprise development would be a step toward this, especially in subsectors such as transport and equipment repair and manufacture that are identifiable as profitable spin-offs from the agro-industrial economy.

BIBLIOGRAPHY

Diaz, A. 1999. *La calidad en el comercio internacional de alimentos.* Lima, Peru, PROMPEX (Comision para la promocion de exportaciones).

Eswaran, M. & Kotwal, A. 1985. A theory of contractual structure in agriculture. *The American Economic Review*, 75(3), 352–367.

Figueroa, A. 1996. Pequeña agricultura y agro-industria en el Perú. *Economía,* XIX: 37–38.

Ministry of Agriculture. 1999. *Producción agrícola 1998.* Lima, Oficina de Información Agraria.

World Bank. 1998. *Peru: an agricultural development strategy.* Washington, DC. (Gray Cover document.)

Chapter 5
Rural farm/non-farm income linkages in northern Ethiopia

Tassew Woldehanna

INTRODUCTION

Rural development policies often neglect the role of rural non-farm activities and their link with agriculture. This might be because the role of the rural non-farm sector is the least understood component of the rural economy; its role in the broader development process is not well known (Lanjouw and Lanjouw, 1997). This knowledge gap has been reflected in the policies of developing countries such as Ethiopia where there is no development policy that identifies and includes the rural non-farm sector. Agricultural ministries have instead focused on farming, and industry ministries have focused on manufacturing. There is a social cost to the failure to recognize the importance of the rural non-farm sector in decreasing rural-urban migration and its potential role in absorbing the growing rural labour force, thus contributing to the national economy and promoting a more equitable distribution of that income.

Despite the recent increase in the literature on farm/non-farm linkages, there has been no significant systematic study conducted on marginal areas in Ethiopia, particularly Tigray. Most studies concentrate on dynamic agriculture, where cash crops are widely grown. The main purpose of this chapter is to inform government agencies, NGOs and donors about the development and constraints of the rural non-farm sector, its link to the farm sector and its importance for rural development policy. This chapter addresses the following questions.

1. What are the patterns and determinants of rural households' non-farm income and participation in non-farm activities?

2. What types of linkages exist between farm and non-farm activities? What is the relative importance of the linkages? Are farm and non-farm income substitutes or complements? What is the influence of non-farm income on income inequality in rural communities?

3. What are the major policy, social, cultural and economic problems for the development of non-farm microenterprises and small enterprises? What kind

of policies, institutional support and technological developments are necessary to alleviate constraints to non-farm employment?

The rest of this chapter is organized as follows: the second section describes the data collected for this analysis and the method. The third section provides an overview of the study region, including the roles of the government and NGOs, the development and constraints of microenterprises in the region and the functioning of the labour market. The fourth section provides empirical evidence on the nature and relative strength of production and consumption linkages in the region, non-farm work participation and its impact, and household-level farm/ non-farm income linkages; the roles of farm and non-farm activities in income inequality among rural households is explored. Policy and programme implications of this research are discussed in the fifth section.

DATA USE AND APPROACH

In order to analyse farm/non-farm linkages in Tigray, a survey of rural households was conducted and there were interviews with labour-market participants and major employers. Secondary data from national and regional statistics offices was also used.

The survey data were collected in two districts, or *woredas*: Enderta and Adigudom in the southern zone of Tigray. The survey consisted of 201 heads of rural households chosen randomly from a stratified sample area. The data include detailed information on the allocation of labour to home, farm and off-farm, income sources, purchase of farm inputs including hired labour, sale of farm outputs, consumption expenditures, credit and household composition. The data was collected for the years 1996 and 1997 for a total of 402 observations.

Data characterizing the surveyed households are given in Table 1. Family size ranges from 1 to 11 people. On average, family size is 5.6 members, which is slightly above the regional average of 4.8 and the national average of 5.2. The average dependency ratio is 58 percent. Most adults are illiterate; only 35 percent of household heads can read and write.

TABLE 1

Description of the household-level data set (n=402)

Variables	Mean
Family size	5.58
Number of dependents	3.26
Land cultivated (*tsimidi*)	7.06
Number of plots cultivated	3.65
Land owned (*tsimidi*)	5.88
Number of plots owned	3.06
Age of the household head	48.00

Note: A *tsimidi* is a local area-measurement unit; 1 ha = 4 *tsimidis*.

Source: Author's calculations from household survey.

Rural households participate in a variety of farm, off-farm and self-employment activities. Farming activities include crop production, accounting for 20 percent of rural households, livestock husbandry, accounting for 6 percent, and mixed farming of crops and livestock, accounting for 69 percent. Traditional farming technology – simple hand tools, ox-driven implements and labour – is the dominant farm input and comes mainly from the family. Only 5.7 percent of rural household use irrigation. Off-farm activities involve wage employment and self-employment. Wage employment includes paid community development work, often called food for work, farm work and manual work in construction, masonry and carpentry. Self-employment includes small trading, transporting goods by pack animal, selling fuelwood, making charcoal, selling fruit, making pottery and handicrafts and stone mining.

BACKGROUND ON THE TIGRAY REGION

Description of the study area

The study area is the Tigray region of northern Ethiopia, which is part of the African dryland zone often called the Sudano-Sahelian region (REST/ NORAGRIC, 1995). The region has an area of 80 000 km² and a population of 3.1 million people in 598 004 households; 85 percent of the population reside in purely rural areas; the remaining 15 percent live either in the regional capital, district centres or rural centres.

Tigray ranges from flat lowland to rugged mountainous plateaux where altitudes range from 500 m in the eastern part, Erob, to 3 900 m in the southern zone near Kisad Kudo. Natural resources are under extreme stress because of the increasing population (REST/NORAGRIC, 1995). Many of the steep slopes have lost their protective cover and are seriously overused by cultivation and livestock grazing. Soil runoff from slopes has caused severe erosion (BPED, 1998). The natural forest of the region has been destroyed, mainly by encroachment of subsistence cultivation; only localized patches of woodland around churches and in remote places remain uncut. The crop production and animal-husbandry potential of the region has declined as a result of degradation of natural resources.

The region does not have well developed infrastructure (BPED, 1998); most areas of Tigray are difficult to reach by mechanized transport. Until 1997, towns in the region did not have a 24-hour supply of electricity; telephone lines and postal services are inadequate and of low quality. Rural radio call telephone systems are found in remote towns in the southern, eastern, western and central zones. There are only seven postal branches in the region.

Table 2 presents data on the level and growth of regional GDP disaggregated by economic activity. At 2 817.66 million *Birr*, the GDP of Tigray constituted 22 percent of the national GDP in 1994/95. Agriculture is the dominant sector at national and regional level. Based on 1995–1996 estimates, agriculture, forestry and fishing account for 64 percent of the regional GDP and 90 percent of employment; industry accounts for 23 percent of regional GDP, services account for 4 percent and other services account for 9 percent. In 1995/96, the overall regional GDP grew by 7.3 percent in real terms. The service sector is the sector with the highest growth in the region at 16.2 percent, followed by the industrial sector at 6.9 percent. The growth of the industrial sector is largely driven by the 124 percent growth in large and medium-scale industry.

Agriculture in Tigray consists of crop husbandry, livestock husbandry and mixed farming, which is the dominant type of farming system. In 1996, for example, the proportion of farm households that were engaged in crop husbandry

TABLE 2

Regional (Tigray) gross domestic product by economic activity at constant factor cost in 1994/95 and 1995/96 (million *Birr* except per capita GDP)

Economic activity	Real gross value		Growth rate (%)
	1994/95	**1995/96**	**1994/95–95/96**
Agriculture, forestry and fishing	1 797.6	1 917.5	6.7
Industry	648.7	693.7	6.9
Mining and quarrying	155.0	175.7	13.4
Manufacturing	73.1	92.3	26.3
large and medium-scale manufacturing	13.2	29.6	124.2
small-scale industry and handicraft	59.9	62.7	4.7
Electricity and water	23.9	25.8	7.7
Construction	396.7	399.8	0.8
Services	371.3	412.5	11.0
Trade, hotel and restaurant	88.0	102.1	16.0
Transport and communication	35.4	41.4	16.7
Other services	247.9	269.1	8.5
Total regional GDP	2 817.7	3 023.7	7.3
Population (million)	3.114	3.193	2.5
Regional per capita GDP (*Birr*)	904.7	946.9	4.7

Source: Regional Bureau of Planning and Economic Development of Tigray Region.

was 19 percent, in livestock husbandry 3 percent and in mixed farming 78 percent. The growing population has decreased the average farm size in the region to only 0.97 ha; 70 percent of farm households own less than 1 ha. Agricultural output is mainly for domestic consumption, although some production is for export, including sesame oil, horse beans, field peas and animal skins and hides.

Farming systems in Tigray are characterized by traditional technology based entirely on animal traction and rainfed land. A variety of crops such as cereals, pulses and oil crops grow in the region. Cereals are the dominant crop; pulses are of secondary importance. The major crops are sorghum, teff, barley and wheat. Vegetables are cultivated in limited amounts in areas where irrigation water is permanent or semi-permanent, and in areas closer to rural centres and urban areas. In Enderta and Adigudom districts, for example, 6 percent of farm households grow vegetables, but 78 percent of them live near towns. Scarcity of arable land leads to extremely intensive land use. Farmlands are owned and run by small farms that are divided into minor plots scattered over large areas. Production is family based, with little hired labour. Livestock except for plough oxen play an important but secondary role. The livestock component increases in the lowland areas, especially in the southern and western zones (REST/NORAGRIC, 1995).

The Tigray region is marginal compared with the southern and central part of the country. Agricultural production in the region is below the national average. In a good year such as 1996, average yield is 1 167 kg/ha nationally but only 1 096 kg/ha in Tigray; in bad years of drought, the yield gap is even greater. The most basic constraints for agricultural development, especially crop production, are unreliable rainfall, lack of oxen, low soil fertility and outbreaks of crop pests. In the central zone, for example, unreliable rainfall is seen by farm households to be the most significant problem, followed by crop pests and lack of oxen (REST/NORAGRIC, 1995). Lack of pasture and fodder are the main constraints in animal production. Scarcity of veterinary clinics is a significant constraint in livestock development. The revival of livestock farming after a drought is very difficult, because a great number of cattle die during droughts.

It will, in short, be very difficult to increase employment in agriculture. Agricultural productivity is very low because of low soil fertility and unreliable rainfall. Livestock husbandry in Tigray is constrained by a shortage of grazing land. Animal dung is used as fuel for cooking, not for enriching the soil; expansion into marginal and steeper slopes is a widespread practice. The result has been widespread degradation of highland soils as a result of erosion. Labour absorption in agriculture is only possible through intensification of agricultural production and the use of irrigation, which is unlikely in the near future. The non-farm sector has not yet sufficiently developed to absorb the growing population; its contribution to the overall employment and income generation remains low.

The role of government and NGOs

The 1974 revolution resulted in a series of policy measures aimed at expanding collective and state-owned farm and non-farm enterprises and managing the economy through central planning. The government restricted individuals to a single type of occupation. Farmers were not allowed to engage in off-farm activities, hire of labour was restricted and farmers were forced to be members of producer and service cooperatives. These cooperatives were given priority for most of the financial-assistance and extension services. Industrial products were distributed through the service cooperatives; private traders in rural areas did not officially exist.

Monopolized public institutions were given the responsibility for promoting the non-farm sector. The Rural Technology Promotion Department (RTPD) was entrusted with the task of developing and promoting improved farm and non-farm technologies and food processing. The Handicraft and Small Industrial Development Agency (HASIDA) was in charge of issuing licenses, organizing cooperatives and assisting in the marketing of products. Ministry of Education adult training centres attempted to teach various handicrafts, construction and farming skills in urban and rural areas. Their efforts were, however, constrained by policy and institutional factors from the very beginning. All promotional activities were directed towards cooperatives. Individuals trained in crafts were unable to establish themselves, because they lacked credit, tools, raw materials and business skills.

When the military government collapsed in 1991, a market-based economy replaced the centrally planned economy. After the formulation of the Federal Democratic Republic of Ethiopia in 1995, the government decided to liberalize the economy and promote investment in the agricultural and industrial sectors. Current government policy emphasizes both sectors, but is less precise with regard to the rural non-farm sector. The focus of the recent economic reform is on structural adjustment aimed at strengthening producers' supply response, developing the private sector, promoting growth of financial intermediation and creating a market for privatization of financial assets.

The main objective of government agricultural policy is to ensure adequate food security through increased agricultural production and employment. A broad based Agricultural Development-led Industrialization (ADLI) strategy has been formulated that concentrates on three priority areas: accelerating growth through the supply of fertilizer, improved seeds and other inputs, expanding small-scale industries to interact with agriculture and increasing exports to pay for the import of capital goods. Under ADLI, a new system of agricultural extension called a participatory demonstration and training extension system was launched in 1994–1995. It provides agricultural inputs in package form with extension advice.

The reform process, particularly structural adjustment, has affected the earlier institutions that were in charge of promoting non-farm activities. RTPD, for instance, was brought under the regional Bureau of Agriculture but has been limited by budget and manpower constraints. HASIDA offers technical and managerial services to small-scale industry and handicrafts, but its operations are financed through the revenue generated by charging for services; it is still being reformed and its services cover only selected urban areas.

With the reform of state institutions, a number of international, national and regional NGOs have become directly involved in farming and non-farming communities. They provide farmers with a variety of services mainly focused on agricultural development, reforestation, soil and water conservation and rural water supply, and credit for income-generating activities, including small-scale trade and handicrafts. The emphasis on rural non-farm activities is minimal. The exceptions are the regional NGOs Relief Society of Tigray (REST) and the Tigray Development Agency (TDA), which are actively engaged in a more diversified set of activities than the international and national NGOs.

REST's involvement in the rural non-farm sector is mainly through its rural credit and saving programme, which operates through 12 branches and 103 sub-branches across the region. It provides loans for cottage and small agro-industries, artisans engaged in rural arts, crafts, horticulture, growing cash crops and rearing pigs. The activities for which loans are provided are:

- crafts such as embroidery, pottery, basket making, spinning, weaving, carpentry and metal work, especially for agricultural implements;
- small-scale trade such as buying and selling at open markets, retailing, hairdressing, tailoring and preparing local food and drugs;
- agriculture such as livestock rearing, bee keeping, horticulture and cereal production.

A loan is provided on a group basis at 12.5 percent interest. The maximum loan is 5 000 *Birr*; the minimum is 50 *Birr*. The duration of the loan is one year, depending on the repayment capacity of the borrower and the nature of the activities. The REST credit programme has improved farmers' access to the financial market, but cannot completely satisfy farm households' demand for credit; loan duration is short and is not coupled with sufficient business advice.

TDA is primarily involved in providing basic education and technical training. It was initially involved in an integrated rural development programme, but since 1996 its focus has been on urban and rural education. TDA runs four technical training centres, two in the central zone in Shire and Axum and two in Mekelle. School dropouts, ex-soldiers, farmers, women and the jobless are allowed to join. Training is provided in basic construction skills such as masonry and carpentry, metalwork, woodwork, electrical systems, motor mechanics and handicrafts such as carpet making. Graduates are provided with the tools and credit to start their own businesses, but their capacity is very limited because of financial and accommodation problems.

In short, it has been and still is unclear which government organization is responsible for the promotion of non-farm activities in rural areas. The minimal promotion of non-farm activities coordinated by the Bureau of Agriculture through RTPD is not suited to rural areas. HASIDA activities are not targeted to rural non-farm activities in general or to rural non-farm activities carried out by farm households. Substantial promotional work on farm and rural non-farm activities has been done by regional NGOs, especially REST and TDA; they seem to be better than the Government at targeting the rural poor and rural non-farm activities, but their activities still require more coordination with government organizations to improve efficiency and avoid duplication.

Micro and small-scale enterprises: developments and constraints

Statistics from 1998 published by the Tigray Regional State Bureau of Industry, Trade and Transport (ITTB) show that small-scale manufacturing enterprises have flourished during the last seven years. In 1991, small-scale industry barely existed except for cottage industries, of which there were 206 in 1994, reflecting remarkable growth. In 1997 they totalled 599, providing employment for approximately five people per establishment. Average capital investment per establishment is 153 000 *Birr*. Grain mills are the most common type of small-scale industry. About 20 percent of small-scale manufacturing enterprises are found in Mekelle, the capital of the region; the remaining 80 percent are found in other zones and *woreda* towns (Table 3). Most of the raw materials used for production are locally produced; imports constitute about 14 percent of total raw materials.

TABLE 3
Distribution of small-scale manufacturing enterprises in Tigray

	Tigray	Mekelle	Southern	East	Central	Western
Number	599	117	117	93	105	168
Investment (*Birr*)	91 651 331	34 252 604	15 550 533	20 872 695	9 369 499	11 983
Employment	2 957	842	765	521	373	558

Source: ITTB Statistical Bulletin No. 1, 1998.

The Central Statistics Authority (CSA) estimates that there are 25 012 cottage and handicraft enterprises, of which 9 percent are found in Mekelle; the remaining 91 percent are in other towns of the region. Cottage industries are known to use more locally produced raw material than small-scale manufacturing industries. Cottage industry in the region covers a variety of industrial groups, including the following major products: food and beverages, textiles and non-metallic mineral goods, which constitute 90 percent of regional cottage industries. Average initial capital invested per establishment for rural areas is 376 *Birr*, and 276 *Birr* for urban areas. Most of the finance for initial investment – 44 percent – comes from people's own savings and from friends and relatives.

Trade, the most common non-farm activity, has grown rapidly over the last seven years. In 1995, the growth rate of this sector was 16 percent. If unlicensed trade by farmer households is included – it is often underestimated in GDP calculations – the growth rate would have been much higher: 234 wholesale businesses, 11 765 retail enterprises and 2 799 service providers have been established in the region, about 16 percent of which are in Mekelle and the rest in zone and *woreda* centres. The initial capital required for the retail trade is lower than for wholesale and service businesses (Table 4). Average initial capital per establishment is 31 301 *Birr* for wholesale, 4 326 *Birr* for retail and 14 992 *Birr* for services. Women own most of the service establishments such as bars and beauty salons, where value-added per unit of investment is the lowest; most wholesale and retail establishments are owned by men. The educational status of the owners of wholesale and retail establishments is comparable; people with elementary education are the owners of most of the trade establishments. The dominant type of ownership is sole proprietorship. In Mekelle, for example, all of the wholesale trade, 98 percent of the retail trade and 99.8 percent of the service trade are under individual proprietorship. A possible reason for the dominance of sole proprietorship is the fear of friction among partners and the transaction costs associated with resolving a dispute.

TABLE 4
Characteristics of commerce in Tigray

Type of trade	Initial capital per establishment (*Birr*)	Female owners (%)	Owner illiterate (%)	Owners grade 1–6 (%)
Wholesale	31 301	11	16	56
Retail	4 326	25	22	61
Service rendering	14 922	71	34	42

Source: Calculated from data provided by ITTB.

TABLE 5
Value-added and employment potential of non-farm activities in Tigray

Type of non-farm activity	Initial capital investment (*Birr*) per unit of employment provided	Value added per unit of investment	Value added per person
Cottage industry	269	2.21	595
Small-scale industry	3 508	1.42	4 966
Total commerce	202 075	0.004	804
Wholesale	156 941	0.04	6 023
Retail	18 000	–	–
Services	426 406	0.004	1 828

Source: Calculated from CSA Statistical Bulletin No. 182.

Cottage industry and small-scale manufacturing industry have a greater role than commerce in providing employment and generating income (Table 5). In Tigray, cottage industries require 269 *Birr* of capital investment to employ one person; the figure for small-scale manufacturing industry is 3 509 *Birr*. Value-added per unit of investment is lower for commerce, generating 0.004 *Birr* per unit of initial investment. Cottage industry generates 2.21 *Birr* per unit of initial investment; the figure for small-scale industry is 1.42 *Birr*.

Constraints to the development of small enterprises and microenterprises fall into two categories: insufficient infrastructure and firm-specific limitations. The infrastructure problem relates to the low quality and insufficient supply of roads, electrical power and telephone lines; the main road connecting other regions and the Government is not well maintained. There was no supply of electricity that could provide power to a manufacturing industry until May 1998 in most urban areas. Since May 1998, most towns have received electricity, but it can still take several months to get electrical power connected because of shortages of electrical equipment. The capacity of the government office responsible for the service is limited. Business people have to spend an enormous amount of time ordering and receiving raw materials and other commodities.

CSA statistical abstracts have documented the firm-specific problems in cottage and small-scale manufacturing industries and trade; these are summarized in Table 6. In cottage handicrafts and small-scale manufacturing enterprises, the major problem is lack of sufficient initial capital, which affects 48 percent of cottage industries 36 percent of small-scale manufacturers. Other problems are lack of adequate start-up skills in cottage industries and lack of raw materials and premises in small-scale enterprises. A few small-scale and cottage industries are not working at full capacity, for which the main reasons are absence of market demand for products, shortages of raw materials and lack of working capital:

TABLE 6

Problems faced by small and microenterprises in Tigray

Business type	Problems in starting business (% of responses)		Operational difficulties (% of responses)	
Cottage/handicraft enterprises	Lack of sufficient initial capital	36	Lack of sufficient initial capital	48
	Lack of continuous supply of raw materials	15	Lack of adequate skill	11
			Absence of market demand	42
	Lack of working premises	12		
Small-scale enterprises	Lack of sufficient initial capital	36	Lack of working capital	16
	Lack of continuous supply of raw materials	15	Shortage of supply of raw materials	11
	Lack of working premises	12		
Wholesale	Lack of sufficient own capital	21	Limited market	29
	Lack of working premises	19	Shortage of working capital	19
	No problem	45	Lack of workplace	9
			No problem	16
Retail	Lack of sufficient own capital	36	Shortage of working capital	38
	Lack of working premises	17	Limited market	31
	Government regulations	6	Lack of workplace	6
Service trade	Lack of working premises	17	Lack of working premises	37
	Lack of sufficient own capital	9	Shortage of working capital	24
	Access to raw materials	7	State harassment	9
	Government regulations	4		

Source: CSA Statistical Bulletin Nos. 172, 179 and 182.

the main problems in trade enterprises are similar. About 6 percent of retail establishments and 4.2 percent of service providers reported that government regulations are a problem in starting businesses; 9 percent of service providers in Tigray report government harassment while operating a business. These problems appear less acute for wholesale traders: about 46 percent of establishments reported no problems in starting up.

Labour markets

Detailed analysis of the labour market is beyond the scope of this paper. This section summarizes the results of an analysis of the labour market conducted using the data collected for the present study (Woldehanna, 1998).

Farm households are not entirely self-sufficient and are partially integrated into the labour market as employers and labourers. The demand for hired labour among smallholders occurs in peak agricultural seasons such as harvesting and weeding and in slack seasons, implying that some farm households are labour constrained even in slack seasons. This shows that public programmes scheduled for slack seasons are not without opportunity costs. The non-farm wage rate for farm households is influenced by location, type of wage employment and year; the rate varies with location, implying that there is a lack of labour mobility, which requires further investigation. Age and education affect the wage rate received by the main woman in the household, but not others. This may be due to the fact that most jobs do not require education (Rosenzweig 1978, 1984 and 1988). The wage rate varies with season and activity, implying that wage rates reflect demand and supply of labour and the amount of effort required to carry out a job. The relevance of subsistence or nutritional determination of wage is thus very low. Wage determination is better explained by the forces of supply and demand and effort required. Farm households in the farm-labour market and other employers in the non-farm labour market rely on

TABLE 7

Forward and backward production linkages in Enderta and Adigudom districts (household-level data)

Linkages	*Birr*/household
Backward production linkages	64.27
Expenditure on fertilizer	61.78
Expenditure on insecticide	0.62
Expenditure on herbicide	0.01
Expenditure on veterinary medicine	1.87
Labour market linkages	
Expenditure on hired farm labour	89.85
Forward production linkages	
Sale of crop output	252.64
Sale of livestock product sold	72.94
Sale of livestock	176.99

Source: Author's calculations from household survey data.

relatives and friends to search for labourers to hire. Most workers rely on relatives and friends to get information about jobs. Labour-market limitations have considerably increased transaction costs associated with hiring labour and searching for jobs. Most of the people working as masons and carpenters acquire their skill after a long apprenticeship, which is slow and unproductive. This has led to a shortage of skilled masons and carpenters for construction and other investment activities.

FARM/NON-FARM INCOME LINKAGES, SURVIVAL STRATEGIES AND INCOME INEQUALITY

Production linkages

Backward and forward production linkages in the region are limited (see Table 7). Farmers purchase few farm inputs such as fertilizer and pesticides. The average value of fertilizer used per household in the southern part of the region is 62 *Birr*, which is a very small percentage of farm output. There is little use of pesticides and veterinary medicine. Households consume most of their farm produce; sales of crop and livestock output are still at a low level. A farm household sells on average only 13 percent of its crop and 15 percent of its production from animal husbandry. Agriculture is in general unable to support major processing industries; there is only one food-processing industry and one tannery in the region.

Table 8 shows the correlation between non-farm activities and a number of variables. The results indicate that district-level non-farm activities are positively related to population density, but its correlation with farm income is very weak. This is because agriculture has limited backward and forward production linkages. The correlation of wholesale and retail trade with agricultural income is much higher than with service trades and small manufacturing industries, which indicates that consumption linkages are higher than production linkages, and that farmers are significant users of wholesale and retail trades. Service trades, small enterprises and microenterprises are negatively correlated with farm output, probably because farmers are forced to participate in non-farm activities when agriculture is unable to support the growing population. This supports the residual sector hypothesis that non-farm activities absorb workers who cannot be readily absorbed into agriculture. Rural centres are an important stimulus for the performance of microenterprises and small-scale enterprises. Districts closer to Mekelle have access to some of the services needed to run small-scale manufacturing industries, such as roads, electricity and telephone lines; the further

TABLE 8

District level correlation between farm income, population density and capital invested in non-farm income in Tigray

	Non-farm activities	Retail trade	Wholesale trade	Service trade	Small manuf. industry
Population density	0.4695	0.5660	0.2614	0.6150	0.2626
Actual farm output (100 kg)	0.0164	0.0904	0.2711	-0.1012	-0.1292
Farm output per capita	0.0430	0.0848	0.3115	-0.0852	-0.1057
Potential farm production	0.0930	0.1345	0.3618	-0.0445	-0.0784
Distance from Mekelle	-0.0931	0.0301	0.1152	-0.1578	-0.1954
Distance from zonal town	-0.1880	-0.1873	-0.0460	-0.2332	-0.1570

Source: Author's calculations.

districts are from Mekelle and rural towns, the lower the amount of capital invested in service trades and small-scale manufacturing.

Consumption linkages

Consumption linkages are the strongest type of linkage in Tigray. Demand for consumption goods increases as agricultural income increases, but the commodity composition of that demand will change as some commodities and services increase in importance while others diminish. Household consumption demands are more complex, with varying income elasticity of demand for individual commodities. Analysis of consumption demand of the farming population therefore deserves special attention.

To analyse the relative importance of different commodity groups in demand linkages, marginal budget shares and expenditure elasticity are derived from Engel functions, estimated using ordinary least squares, which have a nonlinear relationship between consumption and income.

Table 9 summarizes the expenditure behaviour of the average farm household. The results are obtained by evaluating average budget share, marginal budget shares and expenditure elasticity at the sample mean. The commodities consumed are categorized into food and non-food items and into locational groups. The food items include cereals, pulses, oil crops, vegetables and animal products such as milk, butter and cheese, and sugar, tea and salt. The non-food items are grouped into social expenses such as services and ceremonial expenditure, contributions for local organizations and taxes and industrial products such as household durables, clothing and footwear.

TABLE 9
Demand linkages in Enderta and Adigudom districts, southern zone, Tigray

	Average budget share	Marginal budget share	Expenditure elasticity
Total food expenditure	**0.79**	**0.69**	**0.87**
Cereals	0.49	0.20	0.41
Pulses	0.05	0.06	1.20
Oil crops	0.004	0.01	2.79
Vegetables	0.001	0.001	1.00
Animal products	0.11	0.25	2.27
Coffee, sugar, tea, salt, spices	0.14	0.17	1.21
Total non-food	**0.21**	**0.31**	**1.48**
Service, ceremonial and other			
social expenses	0.05	0.11	2.20
Industrial products	0.16	0.20	1.25
Household goods	0.01	0.02	2.00
Clothes, shoes and cosmetics	0.15	0.19	1.27
Locational group			
Own produced food	0.52	0.49	0.94
Purchased food, local	0.13	0.02	0.15
Purchased food, non-local	0.14	0.17	1.21
Industrial products, non-food			
(not locally produced)	0.16	0.20	1.25
Purchased locally, non-food	0.05	0.11	2.20

Source: Author's calculations from household survey data.

Food accounts for 79 percent of total expenditures, leaving a small share of the budget for non-food items. The marginal budget share of food items is 65 percent, which is less than the average budget share. Expenditure elasticity is 0.87, implying that the budget share of food items will decline when total income rises. The result is comparable to the 0.88 reported by Hazel and Hojjati (1995) for Zambia and the 0.81 reported by Hazel and Roell (1983) for the Gusau region of Northern Nigeria.

Among food items, cereals account for 0.49 percent of the budget, but their importance declines as income rises. Expenditure elasticity is 0.41, and the marginal budget share is 0.20, less than half the average budget share. For higher-value food items including pulses, oil crops, vegetables, animal products, and coffee, sugar, tea, salt and spices, expenditure elasticities are very high, which implies that their budget share will increase if household income increases; their average budget shares are currently very low.

All non-food items have high expenditure elasticities, implying that their importance in the budget share will increase as farm-household income rises. The relative increase will be greatest for expenditures in service, ceremonial and other social expenses, and expenditure on clothes and footwear. This clearly shows that agriculture has the potential to strengthen local demand for non-food items in Tigray.

The locational linkage results in Table 9 show that about 86 percent of total expenditure is on regionally produced food and non-food items; the remaining 14 percent are regionally imported non-food items. Expenditure elasticities of imported items, however, are higher than those of local items, which indicates that there are strong household demand linkages to the local economy that are predominantly benefiting the agricultural sector in the short term but that these will diminish when farm-household incomes rise. The average budget share of local products of the non-farm service sector is 0.05; expenditure elasticity is 2.2. Household demand linkages that go to the local non-farm sector such as services and ceremonial expenditures are currently very low, but will increase dramatically when farm-household income rises. Food items imported from outside the region are also potentially important.

Off-farm work participation and its impact on farm income

Rural households in Tigray participate in various off-farm activities such as wage employment and non-farm own business. A summary of rural households' participation in off-farm activities for two districts is shown in Table 10; 81 percent participate in some off-farm activities. Most off-farm work is temporary and does not require skilled labour except for masonry and carpentry. The proportion of households that participate in wage employment is 72 percent; for own business activities the figure is 28 percent. The participation rate in non-farm wage employment is 22 percent. When food for work is excluded, the off-farm work participation rate of farm households is high at 43 percent.

Rural household income can be divided by activity source. On average, farm production accounts for 57 percent of total income; livestock accounts for 16 percent and crop production for 41 percent. Off-farm labour income accounts for 35 percent and non-labour income accounts for 8 percent of total income. The amount of non-labour transfer income for households is very small compared to their farm and off-farm income. The types of transfer income are remittances,

TABLE 10

Farm household participation in off-farm activities in Enderta and Adigudom districts

Type of off-farm activities	Participation rate (%)
Own off-farm business	27.9
Total wage employment	71.5
Non-farm wage employment	21.6
Manual non-farm wage employment	19.2
Masonry and carpentry	3.5
Food for work	57.7
Off-farm work participation excluding food for work	43.0
Overall off-farm work participation	81.0

Source: Author's calculations from household survey data.

47 percent, food aid, 20 percent and gifts and inheritances from relatives, 19 percent.

An economic analysis of the impact of off-farm activities on agricultural productivity and output was conducted in an earlier paper (Woldehanna, 1998). Results showed that income diversification results in higher agricultural output per unit of land. On average, when off-farm income increases by 1 percent, agricultural productivity increases by 0.34 percent. This could be because households learn managerial skills through experience in various activities, reduce soil mining and initiate better farming practices.

Apart from the diversification effect of off-farm activities, off-farm work brings additional income to households that can be used to purchase farm inputs. Results indicate that demand for hired farm labour is highly influenced by farm and off-farm income controlling for other factors such as family size, number of dependents, soil quality, year and location dummies. When farm output increases by 1 percent, expenditure on hired farm labour increases by 0.66 percent; the figure in relation to off-farm output is 0.55 percent. The demand for other variable inputs is also highly influenced by farm output and off-farm income. When farm output increases by 1 percent, expenditure on the variable input increases by 0.94; the figure in relation to off-farm output is 0.43 percent. To sum up the effect of off-farm income on farm productivity through purchase of farm inputs and income diversification, the total elasticity of farm productivity with respect to off-farm income is calculated to be 0.39.

Off-farm income, entry barriers and income inequality

In order to evaluate the relationship between off-farm income and inequality, a Gini decomposition by income source was calculated (Table 11). Crop income has the highest contribution to overall income inequality as measured by the Gini coefficient, followed by wage employment and income from livestock. Crop income, livestock income and wage income decrease income inequality. The results are mixed, however, when wage income is broken down into categories. Paid food-for-work programmes are the only type of off-farm income that decreases income inequality. Non-farm wage and self-employment income increase inequality, as does income from unskilled manual and skilled manual non-farm work. Non-labour income such as gifts, remittances and property rents also increase income inequality.

A possible reason why non-farm income has an unequalizing effect is that there is an entry barrier for the poor. Non-farm wage employment and self-employment require skill and capital to start. In the absence of a perfect credit market, only wealthier households can afford to enter into self-employment. This implies that if rural non-farm activity programmes do not particularly target the poor, wealthy farm households will dominate the most lucrative non-farm activities such as masonry, carpentry and trade. The inequality effect of non-

TABLE 11

Gini decomposition by income sources

Household income components	Mean	S_k	R_k	G_k	G_k*R_k	$\frac{S_k*R_k*}{G_k}$	$(S_k*R_k *G_k)/G$	% elasticity of Gini index
Off-farm self-employment	262	0.068	0.598	0.836	0.500	0.034	0.103	3.5
Off-farm wage	858	0.280	0.489	0.628	0.308	0.086	0.261	-1.9
Food for work	437	0.174	0.183	0.664	0.122	0.021	0.064	-11.0
Manual non-farm wage	284	0.085	0.406	0.883	0.358	0.030	0.092	0.7
Skilled non-farm wage	136	0.022	0.794	0.978	0.777	0.017	0.053	3.0
Non-labour income	194	0.039	0.707	0.951	0.672	0.026	0.080	4.1
Net farm crop income	1 339	0.448	0.698	0.442	0.308	0.138	0.419	-2.9
Livestock income	497	0.164	0.425	0.643	0.273	0.045	0.136	-2.8
Total household income	3 152					0.330		

Note: S_k is the average share of income source k in total income; G_k is the Gini index of inequality for income source k; R_k is the Gini correlation with total income ranking; G is the Gini index of total income inequality; $(S_k*R_k*G_k)/G$ is the percentage contribution of income source k to the Gini index of total income inequality.

Source: Author's calculations from household survey data.

farm income has serious policy implications. If the objective of policy makers is to reduce income inequality, poverty-focused rural non-farm investment has to focus on the type of non-farm activities in which the poor can participate. If this is not possible, the underlying factors that hinder rural households' participation in non-farm activities must be addressed. This requires the establishment of training centres and provision of credit that focus on the rural poor.

DISCUSSION OF POLICY AND PROGRAMME IMPLICATIONS

The need for alternative employment opportunities

It is becoming very difficult to increase regional employment in agriculture. A growing population has decreased farm size, leading to expansion into marginal and steeply sloping land; the result has been widespread degradation of the highlands. Crop residue and animal dung are used as fuel for cooking, not for enriching the soil. Scarcity of land and malaria in lowland areas such as the western zone limit the amount of land under cultivation in the region. Livestock production is not promising either, because forage supplies come from unimproved and overgrazed pasture and crop residue. Poverty is pushing farmers to search for alternative sources of income, particularly wage employment. To reduce the pressure on land, rural non-farm activities have to be expanded. Waiting for non-farm activities to expand until poverty pushes people off the land will further degrade natural resources and eventually increase the cost of rehabilitation. Employing rural people through rural non-farm activities has two advantages: it keeps farmers in the rural areas and reduces rural urban migration, and it provides farmers with additional income and reduces the pressure on land, hence reducing land degradation.

Development of complementary policies and organized promotional activities

Off-farm income is important for the rural economy in Tigray. Rural households with diversified sources of income have higher agricultural productivity. Expenditure on farm inputs is dependent on off-farm income, which helps to finance farming activities, as well as agricultural production. Farmers employed in higher-wage activities such as masonry and carpentry have a greater capacity to hire farm labour. The positive link between farm and off-farm income implies that increasing agricultural output and raising agricultural productivity cannot be done in isolation. Narrowly focused sectoral approaches with the sole target of raising agricultural output and productivity are less likely to achieve significant advances unless considerable attention is given to the importance of non-farm income in the rural economy. Current agricultural extension programmes should include farm and non-farm activities, encourage growth of small-scale business and create non-farm employment opportunities in rural areas. Complementary policies and programmes must be developed to strengthen the link between farm and non-farm activities.

There are attempts in the region to promote rural non-farm activities in order to provide farm households with alternative income sources. Public employment schemes such as food for work, for example, have increased farm households' access to off-farm income to about 35 percent. Efforts to promote off-farm activities are disorganized and insufficient, however, and the links between farm and non-farm activities are not fully recognized. Most government organizations and NGOs have focused exclusively on agriculture, because the majority of the population is engaged in it. Non-farm activities should not be left to the industry and trade ministries; the agricultural ministries should be able to give special focus to rural non-farm activities in order to ensure sustainable farming.

Institutional support might be necessary to create an enabling environment for rural non-farm enterprises. It is not at present clear which government organization is responsible for the promotion of non-farm activities in rural areas, despite efforts by a few NGOs such as REST and TDA. The Bureau of Agriculture concentrates on farming activities and the Industry and Commerce Bureau focuses on non-farm activities in urban centres. A government organization must therefore be established to coordinate promotion of rural non-farm activities; it should be responsible for formulating, upgrading, coordinating and implementing measures such as economic and financial policies to create an enabling environment, and should run assistance programmes to promote rural non-farm activities. The new institution could also lobby for policies that favour rural non-farm activities and development of assistance programmes, because rural non-farm enterprise owners do not have the capacity to organize themselves.

The fact that consumption linkages dominate production linkages signifies that commerce is the main non-farm activity in the short term and should be the focus of government policy. To exploit this potential, government regulations that hinder expansion of business must be avoided. Infrastructure such as roads, electricity and telephone connections must be improved; measures to improve the efficiency of the economy, such as improving the bureaucratic and judiciary systems, would also help. Improving the efficiency of commerce means creating favourable markets for industrial products, especially those of small-scale and cottage manufacturing industries.

Rural towns as a focal point in rural development

Rural towns act as a focal point in the development of the rural non-farm economy. It is essential to ensure adequate economic and social infrastructure to support emerging rural non-farm activities and to renovate and develop traditional ones. Physical infrastructure will undoubtedly play a significant part in strengthening farm/non-farm linkages. Road access to rural towns is essential to stimulate agricultural consumption linkages and provide farm inputs to farm households; the roads will be relatively easy to construct and maintain. Efficient rural credit and smooth labour markets are particularly important to promote rural non-farm activities; investment in human capital is essential to ensure that rural non-farm activities are reliable and able to cope with new technological developments. Most of these service institutions must be located in rural centres, because it would be difficult to establish them in widely scattered settlements.

Targeting of the vulnerable group

Care must be taken in planning programmes to combat rural poverty, because promoting rural non-farm activities will not necessary target the poor. Wealthy farm households dominate the most lucrative non-farm activities, particularly masonry, carpentry and non-farm self-employment such as trade. Poverty-focused rural non-farm investment will need to target non-farm activities in which the poor can participate, or address the underlying factors that prevent poor rural households from participating. This calls for establishment of training centres, provision of credit for the poor, business extension, creation of favourable conditions and improvement of infrastructures.

Women participate to a considerable extent in non-farm activities. They are engaged in activities with lower value-added per unit of investment and low-wage non-farm activities, however, such as public work programmes and manual work in construction sites. Women need to be given training and provided with credit to start their own businesses in order to allow them to participate in well paid rural non-farm activities. To facilitate women's contribution to rural non-farm activities, assistance agencies must be involved and the Government must explicitly recognize the importance of women's roles.

The need to review and update existing policies and institutions

Government policies affect the magnitude of agricultural growth and the ability of rural non-farm enterprises to response to agriculturally induced increases in demand. If rural non-farm enterprises are to achieve their full potential for income generation, policy makers need to review their agricultural, investment and commercial and infrastructure development policies that work against small farmers and small rural non-farm enterprises. A policy that needs reform is the proclamation that provides investment incentives such as income-tax relief to local investors with over 250 000 *Birr* capital, which does not encourage rural non-farm activities that require smaller capital investment. Policies should be formulated to improve access by small rural non-farm enterprises to formal financial institutions such as commercial and development banks.

Searching for jobs and looking for employees are hindered by the fact that most households rely on friends and relatives for information. The transaction cost associated with hiring labour and searching for jobs is exceptionally high, which will retard investment activities and technological innovations in the farm and non-farm sectors. There should therefore be government assistance to help dealers emerge in the labour market in the long term. One way to achieve this would be to cancel the law that prohibits the establishment of dealers in the labour market and to publish in recognized places labour-market information such as wage rates, magnitude and type of labour demand and lists of job seekers by skill until the market supports the emergence of dealers.

Because of the time required to develop masonry and carpentry skills, there is a shortage of well qualified masons and carpenters for construction and other investment activities. Hands-on training of workers in building and construction sites should be organized to improve the performance of investment activities; it would probably be necessary to improve the capacity of government and NGO training institutions and to establish additional training programmes. TDA masonry and carpentry training for farmers should be developed to cover the whole region; vocational schools or local master craftsmen could give the training programme as well. The most important features of successful training programmes are those linked with the labour market: unless training establishments respond to changing labour-market conditions, their graduates will encounter difficulties in finding employment and the investment in training will be socially unproductive.

Choosing the means of intervention

The evidence suggests that programmes aiming to provide a complete package of financial, technical and management assistance are generally less effective than programmes that identify and provide a single missing ingredient such as a small credit programme (Haggblade, Hazell and Brown, 1989). Assistance to firms is usually costly, especially for developing countries, because the rural non-farm enterprises are normally small and geographically dispersed. Direct assistance should focus on system-level opportunities and constraints that open up opportunities for large number of firms (Haggblade, 1995). This kind of highly leveraged intervention requires subsector analysis to identify locations or enterprises supplying inputs or marketing outputs that can expand the income potential for many small firms, upstream or downstream.

Continuous research is necessary to identify opportunities and constraints that hinder the development of rural non-farm activities. National agricultural research centres, universities, international research organizations and donor agencies should include rural non-farm activities in their programmes in order to enhance the effectiveness of the research and increase the familiarity of policy makers and practitioners with the nature and development of rural non-farm enterprises. Surveys by CSA and national and regional bureaux of industry, trade and transport should include rural non-farm activities by large and small firms in the formal and informal sectors.

REFERENCES

BPED. 1998. *Atlas of Tigray*. Mekelle, Tigray, Ethiopia.

CSA. 1997a. Report on cottage/handicraft manufacturing industries Survey. *Statistical Bulletin*, No. 182.

CSA. 1997b. Report on distributive and service trade survey. *Statistical Bulletin*, No. 179.

CSA. 1997c. Report on small-scale manufacturing industries survey. *Statistical Bulletin*, No. 172.

Evans, H.E. & Ngau, P. 1991. Rural/urban relations, household income diversification and agricultural productivity. *Development and Change*, 22: 519–545.

Haggblade, S. 1995. *Promoting rural industrial linkages within agrarian economies for rural poverty alleviation*. Vienna, UNIDO. (Draft report.)

Haggblade, S. & Hazell, P.B.R. 1989. Agricultural technology and farm/non-farm growth linkages. *Agricultural Economics*, 3: 345–364.

Haggblade, S., Hazell, P.B.R. & Brown, J. 1989. Farm/non-farm linkage in rural sub-Saharan Africa. *World Development*, 17(8): 1 173–1 201.

Hazel, P.B.R. & Hojjati, B. 1995. *Farm/non-farm growth linkages in Zambia*. Washington DC, IFPRI/EPTD. (Discussion paper No. 8.)

Hazell, P.B.R. & Röell, A. 1983. *Rural growth linkages: household expenditure patterns in Malaysia and Nigeria*. Washington DC, IFPRI. (Research Report No. 41.)

ITTB. 1998. *Statistical Bulletin*, No. 1, February 1998. Mekelle, Ethiopia.

Lanjouw, J.O. & Lanjouw, P. 1995. Rural non-farm employment: a survey. Washington DC, World Bank. *(Policy Research Working Paper* No.1 463.)

REST/NORAGRIC. 1995. *Farming systems, resource management and household coping strategies in northern Ethiopia: report of a social and agro-ecological baseline study in central Tigray.* Aas, Norway.

Rosenzweig, M.R. 1978. Rural wages, labour supply and land reform: a theoretical and empirical analysis. *American Economic Review*, 68(5): 847–861.

Rosenzweig, M.R. 1984. Determinants of wage rate and labour supply behaviour in the rural sector of a developing country. *In* H.P. Binswanger & M.R. Rosenzweig, eds., *Contractual arrangements, employment and wages in rural labour markets in Asia*, p. 211–241. New Haven, Conn., USA, Yale University Press.

Rosenzweig, M.R. 1988. Labour markets in low-income countries. *In* H. Chenery and T.N. Srinivasan, eds., *Handbook of development economics*, Vol. 1. Amsterdam, Elsevier Science Publisher BV.

Woldehanna, T. 1998. *Farm/non-farm income linkage and the working of the labour market in Tigray, northern Ethiopia.* (Paper for the 23rd Congress of the International Association of Agricultural Economists, Sacramento, Calif., August 1997.)

Chapter 6
Promoting farm/non-farm linkages: a case study of French bean processing in Kenya

Lydia Neema Kimenye

INTRODUCTION

With over 50 percent of the population living below the poverty line, poverty alleviation has necessarily been a primary goal of Kenya's development strategy. Given that over 80 percent of the population and the majority of the poor are based in rural areas, promotion of employment and other income-generating activities in rural areas is crucial to alleviating poverty. The rural areas of Kenya have already been singled out as having "acted as a very effective 'sponge', absorbing and retaining the population and labour, so that rural-urban migration has not been as serious as it could have been, given the high rates of population growth and the limited availability of cultivable land" (World Bank, 1994). What is urgently needed is identification and promotion of economic activities that offer the greatest potential for employment and income generation at grassroots level, so that rural areas can continue to absorb population growth.

Most of the poverty alleviation strategies employed in the past have focused on increased commercialization of smallholder farming, particularly through cultivation of tea and coffee and recently through production of horticultural items for export. Although the farm sector is considered to be the backbone of the economy and a major source of food and income for most Kenyans, recent evidence shows that rural household income in Kenya is increasingly diversified, with a substantial share coming from sources outside the farm. A recent review of several field studies on rural household income diversification in Africa shows the importance of non-farm earnings in rural household income (Reardon, 1997). It is estimated that in Kenya as much as 60 percent of rural household income is gained from non-farm sources (World Bank, 1994); the major sources include non-farm wage employment in rural areas, such as working in agroprocessing enterprises, and profits from small-scale enterprises in the non-farm informal sector.

Evidence from past studies suggests that promotion of farm/non-farm linkages, especially those focusing on commercialization of smallholders, has enormous potential to create employment and to further diversify income. Many smallholders are becoming increasingly commercialized by growing high-value non-traditional crops such as fruits and vegetables for the fresh export and processing markets. Vegetable production is currently the most important commercial horticultural enterprise among smallholders, especially those with very small farms of less than 2 ha. About 80 percent of fresh export vegetables are grown by smallholders, who sell them to exporters through contracts or brokers (Kimenye, 1995). Some of the vegetables are processed and sold in domestic or export markets. Besides providing income directly to farm households that cultivate the crops, commercialization of farm subsectors can generate farm employment and stimulate greater rural non-farm economic activities (Haggblade and Hazell, 1989). Kenyan policy makers should as a matter of importance be informed about the nature of linkages emanating from or towards vibrant smallholder subsectors such as French bean processing, and about ways in which such linkages could be exploited to enhance rural incomes and overall economic growth.

Haggblade and Hazell (1989) have shown that rural non-farm activities across Africa account for about 14 percent of full time employment and between 25 percent and 30 percent of income. Statistics show that in Kenya, at a time when employment opportunities in the public sector are shrinking and the formal sector is growing at only a modest rate, small and microenterprises account for over 50 percent of employment outside agriculture. The informal sector has an annual employment growth rate of over 10 percent (Ministry of Agriculture, Livestock Development and Marketing, 1996). Horticulture, especially fruit and vegetables, maize and dairy commodities were the subsectors identified by government and development agencies as areas with the greatest potential for promoting equity-enhancing small and microenterprises (Ministry of Agriculture, Livestock Development and Marketing, 1996; TechnoServe, 1997). Most microenterprises supply consumers in towns, but it is becoming widely recognized that rural microenterprises are gaining considerable importance across much of Africa as sources of employment and incomes (Jaffee and Morton, 1994) and that they have the potential to generate a variety of linkages within the rural economy.

The objective of this chapter is to characterize the nature of farm/non-farm linkages in Kenya, to explore the ways in which the geographic location of processing firms influences linkage effects, and to identify constraints to the expansion of linkages. These issues need to be addressed in order to enhance employment opportunities and income diversification in rural Kenyan households.

The analysis is based on a detailed case study focusing on the French bean processing subsector, chosen because of its growing importance for the fresh export and processing markets and its potential to benefit smallholders producing French beans and households linked to the subsector. Detailed information is provided on the households that participate in this market and the processing firms.

The remainder of this chapter is organized as follows. Section 2 provides background information for the study and details of the analytical approach. Section 3 examines farm households involved in producing French beans and ways in which production of French beans is linked to the rest of the rural economy. Section 4 investigates the firms processing French beans and their linkages to other sectors. Section 5 contains conclusions.

BACKGROUND

The choice of French beans for analysis of intersectoral linkages is based on the growing importance of horticulture in general and the importance of the French bean in particular. Horticulture is the fastest growing and commercializing agricultural subsector, especially with regard to production for the fresh export and processing markets. Statistics from the Horticultural Crops Development Authority (HCDA) show that exports of fresh horticultural produce grew dramatically from about 1 476 mt in 1968 to 57 363 mt by 1992, worth about KSh2.5 billion or approximately US$48.0 million. In 1996, the export volume of fresh horticultural exports had reached 84 824 mt according to HCDA export statistics. The horticultural processing subsector also recorded a remarkable increase, particularly in terms of processed products destined for the export market. Between 1985 and 1992, for example, the volume of processed horticultural produce rose by 50 percent from 54 465 mt to 81 551 mt worth KSh1.8 billion, or about US$34.0 million. Kenya's processed horticultural products include frozen products, notably French beans, snow peas and juice concentrates, canned products, mainly French beans, baby corn, juices, fruit slices, jams and marmalade, and dehydrated products such as cabbages, carrots, leeks and onions. The dynamism of this sector makes it of special interest and ideal for examination of the development of linkages.

Two types of data were collected for the case study. First, two farm household surveys were conducted, one in Njoro division of Nakuru district and the other in Maragwa division of Maragwa district. The two areas differ in one major respect: in Njoro, processing firms are located in a rural bean-growing area, whereas in Maragwa processing firms are located in Ruaraka, near Nairobi. The household interviews were based on a structured questionnaire. Second, the processing firms supplied by the surveyed households were studied. Informants in the French bean processing firms in Njoro and those based near Nairobi were interviewed through a semi-structured questionnaire.

Maragwa division, situated in Maragwa district in Central Province, covers an area of 20 000 ha; about 15 000 ha of this is suitable for agricultural production. The division has an estimated population of 109 000 people and 17 690 farm families. The average size of farm holding is 1 0.4 ha. In the recently settled Samar and Maragwa Ridge, however, farm sizes reach 80 ha. The main crops grown in Maragwa are maize, beans, bananas, potatoes, coffee, French beans, tomatoes, cabbages and kale; other crops include mangoes, avocadoes, cotton and soybeans. The Ministry of Agriculture estimates that there are currently about 40 ha of land under French beans every season.

Njoro division, a high-potential agricultural area in Nakuru district, covers an area of 774 km^2, all of which is suitable for agricultural production; it has a population of 160 607 people. The size of farm holdings varies from 0.4 ha to 4 000 ha; farms of 8 ha or less are small-scale farms, those between 8 ha and 20 ha are medium-scale and those over 20 ha are large-scale. The major crops are maize, wheat, barley, potatoes, cabbages, kale and pyrethrum. Dairy production is an important enterprise in the division. Large and medium-scale farmers usually have access to irrigated land for production; water usually comes from boreholes. A few smallholders use water from roof catchments to irrigate vegetables. Only a few farmers are producing French beans, because of market limitations.

FARM HOUSEHOLDS AND FRENCH BEAN PRODUCTION

Household characteristics in the study region

In Maragwa, 75 farm households were interviewed; 31 percent of the households in the sample were headed by women. The division has three main trading centres – Maragwa, Gakoigo and Sabasaba – where farm households can obtain production inputs, consumer goods and services, or sell their produce. The distances from farms to the nearest trading centre range from 0.5 km to 10 km, but 72 percent of farmers are within 2 km of a trading centre; Maragwa town is the closest centre for 43 percent of the farmers interviewed.

Most of the French bean growers in Maragwa are young; nearly 60 percent of farmers in the sample are 35 years old or less. According to the Maragwa division agricultural extension officer, French bean production has attracted many of the unemployed young people in the area.

Farm holdings in Maragwa are fairly small: the farmers interviewed had between 0.05 ha and 5.67 ha; average farm size was 0.75 ha. About 5 percent of the respondents did not own land, but rented it to grow beans. Some of the young people grow French beans on portions of land owned by their parents. The survey data indicates that 70 percent of farmers had rented land during 1996; 84 percent of them used it for French bean cultivation. Average hired land per household was 0.18 ha per season; the rental rate was KSh2 400/ha per season; the average amount of land hired per household for the purpose of growing French beans in 1996 was 0.08 ha per season.

The practice of hiring land to grow French beans is more prevalent in Maragwa than in Njoro. Farmers in Maragwa hire additional land to plant French beans, because the amount of land they own is very small and devoted to food crops, and in order to get access to irrigation water. Farmers whose land is at a high elevation and therefore difficult to irrigate often hire portions of land along rivers, on valley floors or near shallow wells for French bean production. The undulating topography of Maragwa has more rivers, valley floors and shallow wells than Njoro. With access to irrigation water, it is possible to grow up to four crops of French beans per year.

In Njoro, 47 farm households were interviewed, of which 51 percent were headed by women. The major trading centre frequented by farmers is Nakuru town, approximately 20 km from Njoro; other local market centres are Njoro town and Egerton centre. Njoro farmers are older than those in Maragwa – 55 percent are over 35 years of age – and have larger farm holdings; average farm size is 1.37 ha. About 79 percent of the farmers in the Njoro sample had hired land from other farmers in the division during 1996; the average size of land hired per household was 0.75 ha. Only 15 percent of the farmers had hired land to plant French beans. The cost of hiring land in Njoro during 1996 was KSh800/ha per season.

French bean production and returns

Two types of French beans are produced in Kenya, one for the fresh export market and the other for processing. Maragwa farmers produce both kinds; Njoro farmers produce only the processing variety. This may be because of Maragwa's proximity to Nairobi, where the fresh-produce exporters operate, and to Jomo Kenyatta International Airport, the main point of export. Maragwa is about 100 km from Nairobi; Njoro is about 250 km away.

Although Maragwa produces both types of bean, the survey data indicates that production for the fresh export market has declined substantially in the last three years, probably as a result of increased competition from the processing subsector and the breakdown of farming contracts between farmers and exporters. Processing firms buy beans from Maragwa under contractual arrangements with smallholders, under which processors usually provide farmers with input credit and farmers are assured of a market. Most fresh-produce exporters do not provide farmers with input credit, however, so farmers are frequently unsure whether a buyer will collect their produce or what the price will be.

The survey indicated that 47 percent of smallholders in Maragwa had been growing French beans for the fresh market in 1994, but that by 1997 only 21 percent were doing so. The average area under French beans for fresh export per farm household has remained roughly the same: 0.08 ha in 1994 and 0.09 ha in 1997.

French bean production for processing has increased substantially in the last three years. Many smallholder farmers, particularly the young or unemployed, have turned to French bean cultivation to earn their livelihood or augment income from traditional crops. Of the 75 farmers randomly sampled in Maragwa, 85 percent were growing French beans for processing in 1997; 67 percent of them started in 1997, 19 percent in 1996, 11 percent in 1995 and 3 percent in 1993. None of the farmers in the sample had been growing French beans for processing five years previously. Two main reasons were given by the farmers as to why they started to grow French beans for processing: about 46 percent started to grow the beans to augment farm income; 44 percent were motivated by the existence of a more reliable market, Frig-O-Ken Ltd. The reliability of the market encouraged 34 percent of the farmers to increase the area planted per season; however, 42 percent of farmers are unable to increase the area under beans because of lack of sufficient labour, especially at harvest, 21 percent have inadequate water for irrigation, and some are not allowed by Frig-O-Ken to increase the area.

Table 1 shows a breakdown of bean growers in Maragwa Division during 1996 by age and farm size. The results confirm the reports of the Maragwa district agricultural office that introduction of French bean production for processing during the early 1990s made farming, and particularly bean cultivation,

TABLE 1

Age and farm size characteristics of French bean growers in Maragwa division, Maragwa district, 1996

Farm household category	Average area of beans per household (ha/ season)	% of farms (n=44)
Age category of farmer		
20–30 years	0.22	39
31–35 years	0.26	17
36–54 years	0.26	32
over 54 years	0.27	12
Total	0.25	100
Farm size category of farmer		
Landless[1]	0.35	8
< 0.4 ha	0.25	41
0.4–1.0 ha	0.23	24
1.0–2.0 ha	0.26	20
> 2.0 ha	–	7
Total	0.25	100

[1] These growers either hired land or borrowed some portion of land from their parents.
Source: Author's calculations from household survey data.

attractive to young people, many of whom turned to it as a source of employment and income. Table 1 indicates that about 39 percent of bean growers in Maragwa are between 20 and 30 years of age; about 41 percent own no more than 0.4 ha of land. In 1996, the mean area under French beans for processing was 0.1 ha per household per cropping season. Of those growing beans, 22 percent were women, who planted an average of 0.06 ha each per season; 78 percent were men, who planted about 0.11 ha each per season.

Production of French beans for processing in Njoro has not increased as remarkably as in Maragwa. Smallholder French bean production had virtually stopped during the time of the survey in 1997. None of the 47 smallholder farmers interviewed were growing French beans in 1997. About 70 percent of the farmers in the sample were growing French beans in 1994, but by 1996 only about 30 percent were doing so. In 1994, the average area under French beans per farm household was 0.18 ha; by 1996 it had declined to only 0.09 ha. Reasons for the decline and eventual disappearance of French bean cultivation include an acute shortage of labour, particularly for harvesting, lack of a reliable market and low prices. As is indicated in the section on farm incomes, growers in Njoro incurred huge losses from the bean crop in 1996, because they lacked a market and because of low prices. A significant proportion of farmers did not harvest their 1996 crop.

The problem of market failure is at first glance perplexing, because there are two processing firms, Njoro Canning Factory and Kokoto Factory, in Njoro division. Njoro Canning Factory is the oldest French bean processing firm in Kenya, but it obtains most of its French beans from contract growers in Vihiga district, about 190 km away, and Kericho district, about 170 km away, primarily because of the availability of water for irrigation in these regions. The factory has its own nucleus farm in Vihiga. Njoro Canning Factory buys 5 percent of its beans from farmers in Njoro Division on an irregular basis.

Kokoto factory was at one time the main market for smallholder farmers in Njoro Division. According to the Njoro divisional agricultural extension officer, a substantial number of smallholder farmers started to grow French beans around 1994 under contract to supply Kokoto factory, which began processing operations in May 1995. The contract growers were tenants on the firm's 23 ha farm; they were supplied with irrigation water from boreholes. In 1996, however, the company experienced a serious cash-flow problem after a large consignment of its processed French beans, worth KSh28.0 million, was rejected by customers in Europe because of poor quality. The firm was therefore unable to pay its contracted growers and suspended the contracts. The Kokoto factory currently produces all of its beans itself.

Production of French beans is highly intensive in terms of labour, fertilizer and agrochemicals. Input credit in kind, and in cash to pay for hired labour at harvest, are critical because the majority of smallholder farmers are resource-poor. In the Maragwa sample, 88 percent of farmers were producing French beans under irrigation; in Njoro, production was rainfed. This is because there are more and cheaper sources of water for irrigation in Maragwa, such as rivers – 85 percent of farmers irrigating French beans get water from rivers – and shallow wells; in Njoro the main source is boreholes. Access to irrigation water enabled farmers in Maragwa to produce up to four crops a year; in Njoro there was only one crop. The most common irrigation system in Maragwa is by bucket: farmers use watering cans to collect the water irrigate the crop manually.

Virtually all smallholder growers in Maragwa sell their produce to Frig-O-Ken under a contract arrangement that specifies the area to be planted per season, planting dates, dates and number of times to spray against pests and diseases, the date to start and finish harvest and the price. Farmers are paid at the end of each season. Although some farmers had formed farmers' self help groups to facilitate negotiation of prices, the final contract is drawn between the individual grower and the company. Under the contract, the grower is supplied with seed, fertilizer and agrochemicals on credit.

During harvest, each grower transports beans daily to the nearest Frig-O-Ken collection point, which is usually about 2 km away. They carry the beans on foot or on bicycles to collection points, where the beans are inspected and weighed; weights are recorded against growers' contract numbers. During 1996, prices offered by Frig-O-Ken were KSh25/kg in the first season, KSh26/kg in the second, KSh27/kg in the third and KSh30/kg in the fourth. In contrast, the prices offered for beans in the fresh-export market showed a higher average over the year, but fluctuated between KSh10/kg and KSh100/kg.

When farmers started to plant French beans in Njoro in 1994–1995, they had contracts with the Kokoto factory. Under the contract, farmers were supplied with input credit in terms of seed, fertilizer and chemicals and each farmer was allocated between 0.1 ha and 0.4 ha on the company farm and supplied with irrigation water. The produce was collected from farmers' plots by the company tractor and taken to the adjacent factory. Under the contract, farmers should be paid at the end of the season, but in 1996 the contracts were suspended or not honoured by the company. Most farmers had no alternative market for the crop. Njoro Canning Factory bought beans from a few farmers, but the price offered was too low and many farmers chose not to harvest. The average price in 1996 was KSh23/kg.

In general, smallholder bean growers seem to be earning significantly lower net returns than at the beginning of the 1990s, despite the current boom in the bean-processing subsector. In 1992, smallholder farmers in Mwea division in Kirinyaga district were earning an average of between KSh6 000/ha per season for non-contract growers and KSh11 200/ha per season for contract growers (Kimenye, 1995). Average net earnings for smallholder farmers in Maragwa in 1996 were KSh2 200 from an average holding of 0.1 ha per year.

Mean gross earnings for bean growers in the sample were KSh14 440 per year; total expenses were KSh8 941, which gave an annual net return of KSh5 499 per farm household. Although low, the net return from French beans was the highest among all farm enterprises. The other major crops were maize, beans and potatoes for food, which showed negative net returns resulting from low yields and low prices. Table 2 shows the distribution of net farm earnings among farmers, based on age and gender. Women farmers obtained higher net returns per year than men, partly because they do a better job of harvesting.

Average gross returns on French beans in the Njoro sample during 1996 were significantly lower than in Maragwa, primarily because most of the growers in Njoro failed to harvest or sell the crop because there was no market. Average net returns per household in Njoro were negative at KSh-2 421. Average expenses were about KSh6 338. Table 3 shows the distribution of net returns for bean growers in Njoro division; the figures indicate that most obtained negative returns from French beans. Unlike smallholder farmers in Maragwa, Njoro farmers have larger holdings and can cultivate maize, wheat, barley and potatoes for commercial purposes and cabbages, carrots and other vegetables on a larger scale for the domestic market.

Marketing problems in terms of lack of a reliable market outlet and low prices are the most serious difficulties confronting growers in the Njoro sample. Lack of water for irrigation and shortage of labour have discouraged Njoro farmers from growing French beans. Farm workers demand higher wage rates for working in the French bean plots, especially to pick the beans. Because of the low price and the unreliable market, many growers were unable to pay higher labour wages and left the crop to seed. Several of the farmers interviewed had a

TABLE 2

Average net returns from French beans, by gender and age group of farmer, Maragwa division, 1996

Farmer category	Net earnings from French beans (KSh/year)
Gender	
Female	6 648
Male	4 991
All farmers	5 499
Age group	
20–30 years	3 269
31–35 years	6 057
36–54 years	7 066
Over 54 years	8 166

Source: Author's calculations from household survey data

TABLE 3

Average net returns from French beans, by gender and age group of farmer, Njoro division, 1996

Farmer category	Net earnings from French beans (KSh/year)
Gender	
Female	175
Male	-5 131
All farmers	-2 421
Age group	
20–30 years	-605
31–35 years	-8 932
36–54 years	-667
Over 54 years	-1 923

Source: Author's calculations from household survey data.

stock of French bean seed, which they did not know what to do with; the seed has a strange taste and is not consumed locally.

The main constraints facing growers in Maragwa Division include lack of labour, insufficient water for irrigation and the high cost of agrochemicals and fertilizer relative to the price they get for beans. A few farmers stated that shortage of land had prevented them from increasing French bean production. Among the problems, the high cost of inputs was stated by 45 percent of farmers to be the most critical constraint, which made for low net returns; 30 percent of farmers stated that their main constraint was was inadequate water for irrigation. Fewer than 10 percent complained about the payments being made at the end of each harvest; they argued that low prices left them unable to hire labour to complete harvests.

Comparison of farm and non-farm income

This section presents a comparison of the contribution of farm and non-farm income to the net income of rural households; the results demonstrate the increasingly important role of the non-farm element. The sources of non-farm income are described, such as wage employment, profit from business and the share of income from French bean production.

Non-farm income is crucial to rural farm households, because it helps to smooth the flow of farm income over the cropping cycle and it stabilizes income by spreading risk through diversification (Lanjouw and Lanjouw, 1995). For smallholders in areas where agricultural output varies greatly over a year or years because of unpredictable weather conditions, the seasonal smoothing and risk diversification obtained through non-farm income sources can be very important.

Table 4 shows the composition of net household income for rural households in Maragwa. The results are fairly consistent with findings from elsewhere showing that non-farm income contributes a significant share to rural household income. The present study estimates that rural households in Maragwa obtain about 63 percent of their income from non-farm sources. The magnitude of non-farm income is small, however. Net farm income was negative mainly because of poor performance of food crops; net returns for French beans were positive.

Sources of non-farm income include wage employment, operating a food kiosk, selling charcoal and retailing maize grain. Only 15 percent of farm households in Maragwa reported owning a non-farm business. The low level of involvement of rural farm households in non-farm business observed in this

TABLE 4

**The composition of net household income for rural households in Maragwa, 1996;
net income per year and proportion of non-farm income in household income**

Farm household category	Net farm income (KSh/year)	Net non-farm income (KSh/year)	% of non-farm income in total
Gender			
Female	-5 092	14 577	25
Male	-9 384	17 792	80
All households	-8 068	16 806	63
Age group			
20–30 years	-12 556	12 289	128
31–35 years	-3 899	12 208	46
36–54 years	-7 329	29 140	9
Over 54 years	-937	3 085	17

Source: Author's calculations from household survey data.

study is consistent with evidence drawn by Reardon (1997) from past studies. Lack of start-up capital was the main reason preventing smallholders from venturing into non-farm enterprises or expanding the ones they already have.

A third of farm households in Maragwa earned non-farm income from wage employment; of these, 72 percent were working as farm workers for their neighbours. The rest of the farmers who were employed off-farm were working in restaurants, food kiosks and canteens, in shops and as civil servants. Of the ten households that reported having a non-farm business, the business belonged to the husband in six cases, and to the wife in four cases. Businesses owned by women were food kiosks, tailoring and charcoal retailing, all within the division; men were involved in the maize trade as middlemen and in selling building sand in Nairobi.

Virtually all the non-farm businesses were started between 1994 and 1996. In general, start-up funds consisted of savings from farm income; none of the owners indicated that the money came strictly from French beans, however. Given the poor performance of other crops, it is nevertheless plausible to assume that income from French bean production could have contributed to start-up funds.

The level of employment generation from non-farm enterprises owned by farm households was very low. Hired labour was used sparingly, because the

TABLE 5

The composition of net household income for rural households in Njoro, 1996; net income per year and proportion of non-farm in household income

Farm household category	Net farm income (KSh/year)	Net non-farm income (KSh/year)	% of non-farm income in total
Gender			
Female	649 606	7 510	1
Male	226 909	17 601	7
All households	442 754	12 448	27
Age group of farmer			
20–30 years	11 365	10 456	48
31–35 years	930 492	9 733	1
36–54 years	600 583	13 326	2
Over 54 years	126 794	16 971	12

Source: Author's calculations from household survey data

enterprises were usually run by the owner with the assistance of one full-time worker and one or two casual workers.

Table 5 shows the composition of rural household income in Njoro, where farm income plays a much larger role than in Maragwa. This may be partly because farm holdings in Njoro are substantially larger than those in Maragwa; there is also greater diversity from more profitable farm enterprises. The proportion of non-farm income in net household income is thus only 27 percent. Although this is low, the estimate obtained for the Njoro sample still falls within the range found across African households (Reardon, 1997). Among farmers between 20 and 30 years old, the proportion of income from non-farm sources is considerably higher at 48 percent, primarily because the majority had off-farm work.

The proportion of farmers owning non-farm business is higher in the Njoro sample than in Maragwa; 23 percent of farm households in Njoro reported that they were self-employed, selling food or general provisions from kiosks in 55 percent of the cases, trading in maize or working as carpenters. All except one belonged to men. With the exception of one maize dealer, all the businesses were located in Njoro Division.

As in Maragwa, non-farm enterprises run by rural households in Njoro do not generate significant rural non-farm employment. The owners run virtually all of them, with assistance from family members on a part-time or full-time basis. Most of the enterprises in Njoro were started between 1994 and 1997. Over 70 percent of the owners said they obtained start-up funds from savings generated from farm income; start-up funds averaged about KSh5 500. French beans were not significant in the generation of start-up capital.

Farm/non-farm linkages at the household level

Anticipated farm/non-farm linkages at the farm level include backward and forward production linkages and consumption linkages. Employment linkages associated with non-farm activities are apparently non-existent because of low participation by smallholders in non-farm enterprises. Because the focus of French bean production at local level is on small-scale producers, production and consumption linkages are especially important in terms of stimulating national economic growth and because they can have profound effects on poverty alleviation and spatial growth patterns.

One of the premises underlying this case study is that the growth of the French bean subsector and the increase in smallholder farmers' production of beans for the processing market had reached a level sufficient to induce supply of non-farm production inputs from within the local area (backward linkages). The hypothesis is that as more and more smallholder farmers cultivate French beans for the processing subsector, a significant proportion of their demand for non-farm inputs such as fertilizers and repair services for farm tools and equipment will be supplied by local sources. The theory on linkages and experience from past studies point to the fact that the type and extent of development of backward linkages is determined by factors such as the farm size, type of crop and agricultural technology, and whether production is rainfed or not. In this study, the size of farm holdings in Maragwa is significantly smaller than those in Njoro division. The technology for producing French beans appears to be the same in both areas, but in Maragwa production is irrigated. These characteristics have in theory different implications regarding the type and magnitude of backward linkages. With large farms in Njoro division, for example, linkages associated with the demand for tractors or ox ploughing and related services would be expected, whereas in Maragwa linkages associated with demand for irrigation equipment would be expected.

As expected, the results show that 51 percent of farm households in Njoro rent tractors from other farmers in the region; none of the farmers in Maragwa reported demand for tractor services or any other farm machinery. None of those who were renting out tractors was captured in our sample, however, perhaps because the farmers who own tractors in Njoro are large-scale farmers who were excluded from the survey. The rental rate for tractors varies from KSh320/ ha to KSh480/ha, depending on whether it is used to plough or to harrow; the mean is KSh400/ha.

In Maragwa, it was found that because production of French beans is irrigated, 87 percent of smallholder farmers had bought at least two or three watering cans; nearly 90 percent of farmers obtained them in Maragwa: 66 percent bought from Maragwa trading centre and 23 percent from Sabasaba trading centre. A local entrepreneur has started manufacturing watering cans in Maragwa trading centre and supplying them to farmers in Maragwa and in nearby Murang'a district. This is an excellent example of how the booming French bean subsector has generated backward linkage into the rural informal non-farm economy.

Another input related to French bean production and the horticulture sector in general is knapsack sprayers, for which there is substantial demand. Farmers use them to spray agrochemicals on to vegetables. In Maragwa, however, the sprayers were used to spray agrochemicals on to cabbages, tomatoes, kale and potatoes, and were rarely used in French bean production. This is because Frig-O-Ken employs people to spray crops. But 46 percent of smallholder farmers in Maragwa purchased knapsack sprayers, over 50 percent locally in Maragwa or Sabasaba; the rest bought them from outside the area, 29 percent from Thika and 17 percent from Nairobi. In Njoro, 57 percent of farmers in the sample purchased knapsack sprayers, mainly from Nakuru town. A few farmers rent sprayers from other farmers in the region.

In terms of backward linkages associated with other traditional production inputs, fewer than 15 percent of the farmers in Maragwa sought production inputs for French bean production from local suppliers, because they obtained supplies from Frig-O-Ken under the terms of the farming contract. Although farmers benefit from input credit given by the processor, the practice limits the spread of backward linkages into the rural small-scale non-farm economy. Other production inputs such as seed maize, dry bean seed, fertilizer and agrochemicals for other crops were obtained locally from a Maragwa trading centre or Sabasaba trading centre. In Njoro, virtually all production inputs used by farmers for major crops during 1996 were obtained from nearby Nakuru, mostly from medium-scale input suppliers. None of the farmers reported buying production inputs in local trading centres such as Njoro, Egerton centre or Pivie market, probably because of the close proximity of Nakuru.

Forward linkages at the farm level are insignificant. The only meaningful forward linkage is the supply of beans to the processors. After production, the beans are transported to the factory or sale point by the farmers, or they are collected from farms by Kokoto factory in Njoro.

In terms of consumption linkages, the results of the case study reflect spending patterns of rural households across Africa, in that a large proportion is spent on food items and less on rurally produced non-food goods such as furniture. This implies that current consumption profiles of rural households do not induce growth linkages into the rural non-farm economy.

FRENCH BEAN PROCESSING FIRMS AND LINKAGES

Evolution of the processing subsector

The first bean-processing firm, founded in the 1970s, was called Njoro Canning Factory; it still exists, but under a different name and management since 1980. It was started as a joint venture between a French company and a local firm. The firm developed a nucleus farm in Vihiga district in what was then Kakamega district and introduced French bean production to smallholders in the area through an outgrower contract-farming scheme. The beans from the company farm and from contracted growers were transported to Njoro for processing and then exported to France. Vihiga is approximately 190 km from Njoro, and its factory still obtains over 50 percent of its raw beans from Vihiga District.

Until the early 1990s, Njoro Canning Factory was the only firm processing French beans for export. Other firms processing horticultural crops focused on fruit products such as juices, slices and jams, and vegetable products such as tomatoes, haricot beans and dehydrated leeks, cabbages, onions and carrots. The information available from export volume and value data and the number and size of processing firms indicates that the turning point for the bean processing subsector occurred in the 1990s. The largest French bean processing firm in Kenya, Frig-O-Ken, was established in 1994. Two other bean-processing firms were established at about the same time: Highland Canners in 1990 and Kokoto Factory in 1995. During the same period, Njoro Canners increased their processing plants at Njoro from one to three; in 1997 the firm established a new French bean canning factory at Juja in Thika district, about 50 km from Nairobi. At the time of the survey in December, 1997 another firm was being set up at Ruiru in Thika district.

By 1997, there were four French bean processing firms in operation. Frig-O-Ken was the largest in terms of estimated volume of beans processed per day, number of permanent employees and casual workers and number of contracted growers; this information is based on 1995 estimates of Frig-O-Ken's operational capacity obtained from field staff and survey data of the three other processing firms. In 1995, Frig-O-Ken had more than 300 full-time employees and over 1 000 casual workers at the factory, compared with 70–100 full-time staff and 100–1 000 casual workers in the other firms. Frig-O-Ken handles over 50 mt of raw beans per day compared with about 2–12 mt per day in the case of the other processing firms (see

TABLE 6

A comparison of the processing capacity of rural farms in Njoro and urban farms in the Nairobi area

Location/name of processing firm	Processing capacity (mt processed/day)
Nairobi/Thika	
Frig-O-Ken (Nairobi)	50
Highland Canners (Nairobi)	3–5[1]
Njoro Canning Factory (Thika)	10
Total	63–65
Njoro	
Njoro Canning Factory (Njoro)	12
Kokoto Factory (Pervie, Njoro)	2
Total	14

[1] Highland Canners used to process about 7 mt of raw materials when the company started processing beans in 1990. The decline in the amount of beans processed results from a problem of supply.
Source: Author interviews.

Table 6). In 1995, Frig-O-Ken had about 3 500 contract growers and was expecting to increase the number to over 8 000 within two years. Information about the firm's current processing situation at the factory and in the field was unfortunately not available, because the company declined to be interviewed.

Among the main reasons for the proliferation of bean-processing firms in and around Nairobi, an urban area where beans were not previously grown, are that the city is close to the main French bean growing areas of Murang'a, Maragwa, Thika, Kiambu, Kirinyaga and Nyeri in Central Province and Embu, Meru, Machakos and Makueni in Eastern Province, and close to other firms that supply important secondary inputs used in food processing, such as containers for raw and finished products, and firms that supply equipment and spare parts. Bean processing firms use large quantities of cardboard cartons and plastic crates when transporting French beans from farms to the processing factories and during sorting, grading and handling of the beans prior to processing. Other inputs used are glass jars and cans for packaging the finished products. Most of the firms that manufacture handling and packaging materials are found in Nairobi and Thika town. Metal Box Kenya, for example, is the largest manufacture of cans in Kenya; its head office is in Nairobi and the factory is in Thika town. Most food processing firms that use cans for packaging of the finished products obtain their cans from the Metal Box factory. Proximity to Nairobi is also convenient for the firms in terms of following and obtaining quick administration of official matters such as renewal of licences and permits, exemptions or payments of duty and resolving income tax issues. It is often necessary in Kenya to follow up official bureaucratic processes personally in order to speed up the paperwork. Being closer to the government offices can make a difference in terms of time and money. Road conditionss in Kenya are very poor, so Nairobi's central location and direct access to the main gateways for export of horticultural produce – Jomo Kenyatta international airport and the port of Mombasa – make it a convenient location for the business. Some of the produce, especially frozen French beans, are transported in refrigerated trucks or containers to the export exit points.

The nature of linkages associated with the processing firms

The emergence of French bean processing firms has stimulated local smallholder cultivation. Contrary to the expectation from theory whereby a firm invests in a processing factory or increases its capacity in response to increased production of raw material that has attained a certain critical mass, experience in Kenya suggests that bean firms emerged in response to an export demand for processed beans. In turn, the bean firms established their own farms in order to supply some of the beans and then bought the rest from smallholders through contract-farming arrangements. The establishment of contract farming that often includes provision of input credit is major factor in motivating smallholders, especially those with very small farms, to concentrate on French beans for the processing market. In spite of the low levels of net return from French beans because of the low prices offered by the processors, the enterprise is still the major source of income for smallholders in a region with high population density and without alternative cash crops such as tea or coffee.

The data from the survey of bean firms indicate that in general the processing firms obtain between 50 percent and 75 percent of raw materials from smallholders, mainly in the central, eastern, western and Rift Valley provinces. Njoro Canning Factory obtains 50 percent of the beans it processes from its own farm, 45 percent from contract growers and 5 percent from non-contract growers. The number of contract growers is currently about 200 in Vihiga and 30 in Kericho. Highland Canners gets 25 percent from its own farm of about 137 ha in Embu district and 75 percent from smallholders in Murang'a, Maragwa, Mwea and Thika districts.

Bean processing firms generate backward production linkages to traditional production-input dealers. Virtually all the processing firms own a nucleus farm to produce a proportion of their beans. Where the processing firm has established farming contracts with smallholder bean growers, the firm supplies farmers with production inputs in credit form. All the firms stated that they buy inputs such as fertilizer and agrochemicals in bulk from dealers, in Nakuru in the case of Njoro Canning Factory and Kokoto Factory and Nairobi in the case of Highland Canners. This means that none of the business associated with the provision of production inputs is injected into the local community.

Another production linkage deals with transport of beans from production areas to processing factories. The available information suggests that transport of raw materials from farms to the processing factories does not offer significant opportunities for development of linkages between the bean firms and truckers. Only Highland Canners reported use of middlemen for transport of beans sourced from smallholders. This may be because Highland Canners does not have farming contracts with smallholders and relies on middlemen to buy beans from farmers. Highland Canners uses its own trucks to transport beans from its own farm, accounting for 25 percent of the beans it processes; the rest are transported by middlemen locally referred to as dealers or brokers. The firm uses three to five dealers per week during peak seasons.

The other aspect of the transport linkage is moving finished products from the factory to the point of export. The available data indicates that development of this type of linkage is limited because most bean processing firms use their own trucks for transport to the point of export. This is because the products, particularly frozen beans, are often transported in special refrigerated trucks owned by the firms. Njoro Canning Factory indicated that about 20 percent of its finished product is transported to Mombasa in hired transport. Every week in the peak season, about ten 20-ton refrigerated container trucks transport produce to Mombasa; two or three of them are hired.

The case study identified other important auxiliary activities which arise indirectly from or to the bean-processing firms, including activities such as providing special secondary inputs such as packaging and services such as repair and maintenance of factory equipment and vehicles, clearing and forwarding and legal and medical services. Most bean firms employ a resident mechanic and have a workshop for minor maintenance work, but they also use servicing firms from nearby towns. The firms contract dealers, usually from Nairobi, to conduct annual safety inspections of the plants.

The most promising spin-off activity is provision of packaging materials, for example glass jars, cans, plastic crates and cardboard cartons, and uniforms for factory workers. Processing firms used to import the glass jars, but they can now be obtained locally in Nairobi. The data from Njoro Canning Factory, which was the most complete in terms of annual demand for packaging materials, indicate that a medium-scale processing firm would have the following annual requirements for packaging materials:

- about 1 300 plastic crates every 15 months bought from a manufacturer in Nairobi; at the price of about KSh100 per item in 1997, this amounts to KSh1.3 million;

- between 150 000 and 300 000 cans every month, obtained from Nairobi or Thika and costing between KSh2.4million and KSh4.8 million every month at peak season;

- uniforms for about 1 000 factory workers, two complete sets per year obtained at the closest major town, usually consisting of dust coat, cap and gum boots for the canning factory and pullovers, woollen caps and gum boots for the freezing plant.

Other spin-off linkages include supply of medical care to permanent employees and occasional medical service to casual workers in the case of factory accidents. Njoro Canning Factory hires the services of a private clinic in Njoro to treat employees. A spin-off service identified in Njoro and Nairobi was provision of lunch and tea for factory workers. At Njoro Canning Factory, the management invited a local resident to set up a canteen in the factory where lunches and teas are served. Under an arrangement between the management and the canteen operator, the workers can have their meals at the canteen on credit, with the amount consumed deducted from wages. In Nairobi, women come to the factory gates to sell hot food, usually a mixture of maize and beans, and porridge, tea and fruit, usually bananas.

Table 7 shows the type and amount of direct employment generated by the Kokoto factory at Pivie in Njoro division, a small-scale processing factory with a capacity of about 2 mt of raw beans per day at peak seasons. It obtains beans from its own adjacent farm; apart from the senior managers, virtually all its factory and field employees come from the community surrounding the factory. Some of its casual workers come from families that used to grow beans for the company in 1995, before the contractual arrangements with smallholders broke down. Kokoto Factory is about 15 km from Njoro Canning Factory.

TABLE 7
Type and amount of employment from a small-scale bean-processing firm in a rural area

	No. employed at factory		No. employed in the field	
Gender	Permanent	Casual	Permanent	Casual
Female	–	–	3	92
Male	–	–	4	50
Total	20	70	7	142

Source: Author interviews.

As shown in Table 7, a single small-scale processing firm such as Kokoto Factory does not generate much direct employment; nearly 60 percent of those employed are casual labour, working in on-farm jobs. Of the work generated on-farm, 65 percent is for women, because more labour is required for harvesting than for other activities; women are preferred to men because of their gentle handling of the beans. The data obtained does not include a gender breakdown of those employed at the factory.

Although small, the number of workers employed in the field increased in the three years following establishment of the firm. In 1995, the firm had 2 permanent and 70 casual employees working in the field, advising and assisting the contract growers. At the time of writing it had 7 permanent employees and 142 casual workers, an increase of over 100 percent. Such growth in employment would be significant if there were more firms in the area.

Direct on-farm employment potential by other processing firms, including the Njoro Canning Factory and Frig-O-Ken, cannot be determined because the field production manager for Njoro Canning Factory was on leave at the time of the survey and Frig-O-Ken refused to be interviewed. Evidence from Kokoto Factory regarding the type and number of workers and the growth rate in employment over the last three years indicate that the direct on-farm employment creation potential for a medium-scale bean-processing firm with a nucleus farm and smallholder contracts is significant.

The data used to show direct non-farm employment generation potential and to illustrate the employment leakage resulting from the location of processing firms outside the rural bean-producing areas are drawn from information collected from Njoro Canning Factory and Highland Canners. Njoro Canning Factory is not located in a bean-producing area. Table 8 shows a breakdown of the type and amount of labour employed at Njoro Canning Factory and Highland Canners.

TABLE 8

Type and amount of employment from medium-sized bean-processing firms in a non-bean-producing area

	No. employed at Njoro canning factory (full capacity; excluding Juja factory)			No. employed at Highland Canners (full capacity)	
Gender	Permanent	Casual	Contract	Permanent	Casual
Female	50	800	15	60	100
Male	50	200	5	10	5
Total	100	1 000	20	70	105

Source: Author interviews.

A large percentage of the employment generated is for women casual workers: 87 percent of the employees at the factories are casual and 79 percent are women. The processing firms prefer casual labour because they do not operate at full capacity throughout the year. Women are preferred to men at the factory because most of the work is delicate, such as nipping off the tips of the beans and packing. At Njoro Canning Factory, about 50 percent of factory employees come from the surrounding community within 10 km of the factory. Because it is in a rural area, the factory contributes to reducing rural unemployment. The workers at Highland Canners in Nairobi are from the city and the nearby town of Thika.

Of the three processing firms interviewed, two indicated that they had reinvested some of their profits in the local area. Kokoto Factory reinvested some of its profits locally by establishing a bakery that supplies bread in Njoro and Nakuru and has diversified its activities into fruit processing; the firm processes about 3 000 pineapples per day and the produce is exported. Njoro Canning Factory has made two major reinvestments since 1980. It recently invested in a medium-scale processing factory at Juja in Thika district. The new factory is situated approximately 300 kilometres away, however, and so has no impact on employment and income of the residents in Njoro Division. The other reinvestment was the purchase of the Pan-African Vegetable Dehydration Plant at Naivasha and its re-establishment in Njoro town. Although highly capital-intensive and therefore generating only a small amount of non-farm employment, the dehydration plant has benefited local farmers because it is an alternative market for their cabbages, carrots, onions and leeks. In 1996, the company lost its major European market, Germany, when one of the German partners in the dehydration business left the company. The firm then dehydrated vegetables on a very small scale for the domestic market, mainly East Africa Industries, which uses the dehydrated and ground vegetables as a base for Royco Muchuzi Mix, a popular spice in the Kenyan market. The company identified another client in the export market, however, and resumed large-scale vegetable dehydration in 1998.

The extent to which the processing firms develop or promote linkages that enhance the employment and income of rural people is not related to their locations: neither the Nairobi-based nor the Njoro-based firm has generated significant on-farm linkages with rural bean-producing areas. Virtually all the non-farm growth linkages generate auxiliary business in large towns, particularly Nairobi. This is because the type of linkage activities that generate the most lucrative business seem to be capital-intensive, such as operating refrigerated container trucks and producing glass jars and cans. These capital-intensive, large-scale input supplying enterprises are owned by big firms or dealers in large urban areas.

Factors that have constrained or enhanced the growth of processing firms

None of the French bean processing firms cited any form of government regulation or policy on horticultural production, processing and marketing that has constrained them. The government has in fact played a significant role as a facilitator for the private sector, recognizing the immense contribution of the horticultural sector in generating employment, income and foreign exchange. HCDA provides advisory services to farmers, processors and exporters, particularly in marketing extension and initiation and organization of farmers' groups to meet the needs of processors and exporters.

Another factor that has enhanced the growth of the French bean subsector is the ready export market for processed beans. Virtually all the firms interviewed, however, stated that identification of clients in the export market was easier if the firm had a partner or associate from the destination market who could ensure that quality requirements are met. All the firms visited had a manager from Europe, usually France or Germany, in charge of quality control at the factory.

Among the factors that have severely constrained the subsector is the poor and deteriorating condition of the highway from Nairobi to Mombasa and the roads in rural bean growing areas and urban centres, including Nairobi. The problem of poor roads needs to be addressed urgently if the growth rates of the processing subsector and the whole horticultural sector are to be sustained.

All the firms visited complained of problems of insufficient supply of water for irrigation, which made supply of raw materials less dependable, especially from smallholders. With development of more sources of irrigation water, such as boreholes, production could be increased in the current bean-growing areas and new areas could be opened up for growing and possibly local processing.

CONCLUSIONS AND RECOMMENDATIONS

The French bean processing subsector is one of the most rapidly growing subsectors, after cut flowers, in Kenya's agricultural sector. The case study revealed that smallholders, particularly those in densely populated areas with very small farms, rely on French bean production as a major income-generating enterprise. The results indicate that many young unemployed people consider French bean cultivation as a source of employment, income and livelihood. Smallholders in the densely populated Maragwa division, whose farm holdings average 0.4 ha, have rapidly adopted French bean production in recent years.

A major factor in the rapid adoption of bean production in Maragwa is the availability of an accessible and reliable market outlet, the processing market. Although it is based in Nairobi, the processing firm has established French bean buying centres in the area. All growers have farming contracts with the processor, which assures them of a market and input credit.

The results of the study imply that the growth of the French bean processing subsector over the last three years has not contributed significantly to the development and spread of farm/non-farm linkages in the rural economy. It made no difference whether the processing firms were located in rural or urban areas: the effect on the development of non-farm linkages in the rural areas was insignificant in both situations. Most of the non-farm linkages generated as a result of growth in the subsector were experienced in the urban area, particularly Nairobi and its environs and to a limited extent in Nakuru town. These linkage activities stemmed largely from the spin-off linkage activities, particularly auxiliary firms that supply the agro-industry with packaging materials and services.

Another potential avenue for development of non-farm linkages in rural areas is backward production linkages related to the supply of production inputs such as agrochemicals and fertilizers. The demand for these inputs, particularly in Maragwa, appears to be high and probably capable of supporting informal small-scale suppliers in the area. The processing firms procure these inputs from Nairobi, however, and distribute them to smallholders on credit.

The poor condition of Kenyan roads is a major factor contributing to the firms' preference for large urban areas such as Nairobi in terms of setting up their factories and associating with urban based input and service suppliers. Bad roads in most rural areas increase transportation costs and introduce high risks of failing to deliver supplies and materials on time. This is an issue that needs to be urgently addressed by policy-makers.

The findings of the study show the role of irrigation in increasing output and incomes and in terms of stimulating non-farm linkages in rural areas. It was found that in Maragwa, where French beans are produced under irrigation, farmers are able to grow four crops per year; in Njoro, where production is largely rainfed, growers can only produce one crop per year. Because of the high demand for watering cans that resulted from the increase in irrigated French bean production in Maragwa, an informal rural non-farm business has emerged to supply the cans from within the area. The availability of irrigated production also influences the processing firms selection of the regions from which to procure their raw materials. It is therefore crucial that efforts to encourage smallholders to produce French beans or other horticultural crops for export and processing markets, as currently being tried in parts of Rift Valley Province, should seriously investigate the availability of water for irrigation and determine ways to improve water access for smallholders in a cost-effective way.

REFERENCES

Haggblade, S. & Hazell, P. 1989. Agricultural technology and farm/non-farm growth linkages. *Agricultural Economics*, 3: 345–364.

Jaffee, S. & Morton, J. 1994. *Africa's agro-entrepreneurs: private-sector processing and marketing of high-value foods.* Washington DC, World Bank Environmentally Sustainable Development Division. (AFTES Working Paper No.15.)

Kimenye, L.N. 1995. Kenya's experience in promoting smallholder production of flowers and vegetables for European markets. *African Rural and Urban Studies*, 2(2/3): 121–141.

Lanjouw, J.O. & Lanjouw, P. 1995. *Rural non-farm employment: a survey.* Washington DC, World Bank. (Policy Research Working Paper No. 1 463.)

Ministry of Agriculture, Livestock Development and Marketing. 1996. *Kenya: agriculture strategy for agricultural growth.* Nairobi.

Reardon, T. 1997. Using evidence of household income diversification to inform study of the rural non-farm labour market in Africa. *World Development*, 25(5): 735–747.

TechnoServe. 1997. *Delivery of non-financial services to micro and small enterprises in Kenya.* Nairobi. (Paper for workshop at Barclays Bank Staff Training Centre, Karen, Nairobi, Kenya. May.)

World Bank. 1994. *Kenya: employment growth for poverty alleviation.* Washington DC.

Chapter 7
The potential for farm/non-farm linkages in the cassava subsector in Ghana

Ramatu Al-Hassan and Irene Egyir

INTRODUCTION

This chapter focuses on the cassava subsector in Ghana. The choice of cassava for a study on farm/non-farm linkages was made because of its growing importance as a crop in Ghana and because it is almost always processed in some form for sale. The need to process cassava enhances its potential as a commodity linked to the non-farm sector. In recent years, there has been an increase in the market for cassava chips for export. Although the chip market has become more uncertain in recent times, it continues to influence the market for traditionally processed cassava products and has fostered a number of linkages between the farm and non-farm sectors.

The analysis of the evolving market for cassava products in this chapter is organized as follows. In the second section, background information on farm/ non-farm linkages and cassava production in Ghana is presented. In the third section, information is given about the two districts that form the basis of this study. Linkages between the cassava market and the rest of the economy are examined in the fourth section; this is done by examining each of the markets for processed cassava. The fifth section examines the recent evolution of the cassava market and how this evolution has affected rural producers and the linkages within the market. The sixth section examines the role that institutions have played in forming linkages and supporting cassava production, particularly the Government. The final section presents conclusions.

CONCEPTUAL FRAMEWORK AND BACKGROUND

The weak farm/non-farm linkages in the Ghanaian economy have constituted one of the major development issues since independence. In a report on growth and poverty in Ghana, for example, the World Bank noted the following: "Unless substantial improvements are made in the performance of agriculture, the potential for growth elsewhere, particularly in agro-industries and transport, could be restricted because of intersectoral linkages." (World Bank, 1993). Yet there has been very little study of linkages in the Ghanaian economy; the research has to tended to focus on the macro level, using input-output analyses (Stryker and Dumenu, 1986; Jebuni, Asuming-Brempong and Fosu, 1990); these studies identify weak linkages between agriculture and the industrial sector. The Plan Consult report, for example notes that intersectoral linkages in Ghanaian agriculture amount to only 13 percent of cross-sectoral activities, and that even though agriculture employs 55 percent of the active labour force, its contribution to gross domestic product is only 45 percent. Jebuni, Asuming-Brempong and Fosu (1990) report that early input-output tables on the Ghanaian economy show that low use of intermediate inputs in the agricultural sector is a major reason for the weak production linkages between agriculture and industry. The intermediate input demand of the agricultural sector, excluding cocoa, forestry and fishing, was estimated at 5 percent of agricultural output; intermediate input use of agricultural output by the rest of the economy was estimated at 7 percent. Demand for agricultural output as an intermediate input came mainly from food and beverage processing, tobacco and textiles. Similar conclusions have been reached by other studies, which estimate that intermediate inputs, mainly seed and hand tools, account for less than 10 percent of the value of cocoa and less than 20 percent of the value of maize and rice (Stryker and Dumenu, 1986).

To assess demand linkages, Jebuni, Asuming-Brempong and Fosu (1990) estimated an econometric relationship between non-farm expenditures and an index of agricultural output, agricultural terms of trade, food terms of trade and non-farm output. Their results show that a unit increase in agricultural output induces an increase of 0.52 in demand for non-farm output; a unit change in agriculture's terms of trade stimulates a unit change of 0.49 in non-farm consumption expenditure. Unlike the weak production linkages, these estimates suggest fairly strong demand linkages between agriculture and the non-farm sector. To examine this further, Table 1 presents data on expenditure shares by income class. In general, households spend over half of their budget on food items, although poorer households use more home-produced food than richer households. Richer households spend a smaller share on food than poorer households, but spend a greater share on non-farm items. Although a marginal budget share analysis is required to give more conclusive results of the linkage implications of the expenditure patterns, the budget profile suggests that increases in income levels will lead to greater non-farm demand.

Despite high shares of food in the budget shares of all income classes, the domestic market for agricultural output remains rather small because of the small population and the low purchasing power of domestic consumers. Nearly

TABLE 1
Expenditure shares by income class, 1987/88 and 1988/89

Expenditure type	Extremely poor		Moderately poor		Non-poor		Rich	
	(a)	(b)	(a)	(b)	(a)	(b)	(a)	(b)
Home food	33	32	32	28	24	17	13	7
Food bought	33	35	35	37	39	40	40	38
Home non-farm	3	3	4	4	7	8	13	15
Non-farm bought	23	23	22	23	23	26.	24	27
Other	7	7	7	8	7	10	10	14

(a) 1987/88 Ghana Living Standard Surveys (GLSS) data; (b) 1988/89 GLSS data.
Source: Seini, Nyanteng and van den Boom (1997).

50 percent of households had expenditures below the minimum wage between 1988 and 1989; as a result, the Ministry of Agriculture has sought an export-driven growth strategy for agriculture since the late 1980s (Ministry of Agriculture, 1990). Such a strategy needs to be accompanied by product development to diversify markets and increased opportunities for value-added to agricultural products, whether for domestic markets or export. The activities associated with export diversification, product development and processing are expected to create the additional demand for agricultural output and generate additional employment in the sector, thereby raising incomes and spurring further demand for agricultural and non-farm consumer goods and services.

The cassava subsector has the potential to generate this type of growth. Cassava has for a long time been grown by smallholders as a food-security crop because of its resilience to adverse climatic conditions and its low input requirements. Production of the crop has grown steadily, mainly by area expansion through replacement of fallow and as a substitute for other crops such as yams. Average yield per hectare has improved from 7 mt in the early 1980s to about 12 mt in the 1990s. Ministry of Food and Agriculture statistics show that since 1994 Ghana has produced about 1 million tons of cassava as excess supply (Ministry of Food and Agriculture, 1996). Cassava has potential as an export commodity because of world market demand for cassava as animal feed. Cassava chip exports increased from 500 mt in 1993 to over 20 000 mt in 1997 (Trade and Commodity Group [TCG], personal communication, 1996). These figures exclude exports of *gari*, cassava dough and tapioca, which is granulated cassava starch, as ethnic foods to Europe and America, and through informal cross-border trade, particularly Togo. Actual exports of cassava products are likely to be much higher than reported.

Cassava has a high potential for product diversification, because it can be processed into various forms for human consumption and made into chips for animal feed; starch and starch derivatives can be extracted for industrial uses. Cassava utilization in Ghana is still mainly for human consumption, however, either as unprocessed roots or in the traditional processed forms of *gari*, dough and flour. The bulk and perishability of the crop mean that demand for transport is high. These processing and transport requirements suggest that cassava will have substantial forward linkages.

The cassava subsector has been increasing production over the last few decades. The present estimated yield of 12 mt/ha is still well below the achievable yields of 30 mt. Recent yield increases suggest that the cassava production system is being transformed from an extensive subsistence system to an intensive commercial system. If this is occurring, it is worthwhile determining the factors driving the change and the input demand being created from it.

THE STUDY AREA

The study was conducted in Nkwanta district in the Volta region and Atebubu district in the Brong Ahafo region, which were selected for the study because they are high-potential agricultural areas and currently involved in processing cassava chips for export as animal feed. They are therefore well placed for an assessment of the effects on the farm and non-farm sectors of an increase in demand for cassava. A brief description of the economies of the two districts is given below.

Atebubu district

Atebubu district has an area of 6 720 km²; the population increased from 50 225 in 1960 to 141 181 in 1996, at growth rates of 2 percent between 1960 and 1970, 5.4 percent between 1970 and 1984 and 4 percent between 1984 and 1995. The population is 60 percent rural; the remaining inhabitants are distributed among towns of 5 000 or more people situated along the main road linking the population centres of Yeji and Kumasi. Population density is only 21 persons per km². The district experiences low out-migration of 14 per 1 000, but high in-migration from the north and south. Migrants from the north come mainly for farming; those from the south come to trade in agricultural produce, especially yams. The Atebubu district is in the forest/savannah transitional zone, where there is bimodal rainfall with peaks during May and June and during September and October. Mean annual rainfall ranges from 125 cm to 140 cm. Atebubu serves as a transit point between north and south; the towns of Atebubu and Yeji play important roles, although traffic through the district has diminished drastically since the construction of the Kumasi-Tamale road via Kintampo.

Soils in the district support rice, yams, cassava, maize, sorghum, groundnuts, soya beans, cowpeas, cotton, kenaf, vegetables and tobacco. The bulk of agricultural produce is sold in the production areas. The second most important sale point is the weekly district market. Market interaction outside the regions is limited, probably because the roads are bad and farmers are unable to move produce themselves; traders are able to use tractors to transport produce from buying centres.

Agriculture is the dominant sector of the economy, occupying 60 percent of the population. It is estimated that crop production generates 65 percent of the district's income. Expenditure shares are 42 percent on food and 23 percent on farming, suggesting that the non-farm sector is small and its components classified as:

- agriculture-based: milling, brewing, fishing, *gari* processing and distilling;

- wood-based: carpentry and charcoal burning;

- textile-based: tailoring and dressmaking;

- clay-based: pottery;

- metal-based: ironwork.

Agriculture-based activities account for 37 percent of the non-farm sector, followed by wood-based activities with 33 percent.

The district is poorly served with infrastructure. The only major road runs through the district capital from Kumasi in the south to Yeji in the north. There is no piped-water network in the district. Five banks serve the district, including two Ghana Commercial Banks, one Agricultural Development Bank and two Rural Banks; the main beneficiaries of bank services are salaried workers and traders. Farmers tend not to save in banks, and loans granted to farmers have been difficult to recover.

Nkwanta district

The Nkwanta district is the largest district in the Volta region, with an area of 4 530 km^2. The projected population in 1995 was 150 000 with a density of 33 people per km^2, which is sparse compared with the regional density of 70 persons per km^2 Processing of different types of cassava products is associated with ethnic groups, which include the Konkomba, Ntwumuru, Challas, Atwode, Ntrubu and Adele. The district stretches from the savannah woodlands of the north to the forest zone in the south. Mean annual rainfall is bimodal and ranges from 130 cm to 170 cm.

The various agro-ecologies in the district offer good growing conditions for a wide range of crops. The main crops grown in the savannah ochrosols include yams, cassava, maize, groundnuts, cowpeas and sorghum. The associated bottom soils in the river valleys are used for rice production. The more fertile forest ochrosols and oxysols are planted with cocoa, citrus, oilpalm, cola, plantain, bananas, cocoyams and pineapples.

The economy of the district is largely agricultural; the Department of Agricultural Extension estimates that more than 80 percent of the population is engaged in agriculture, although this estimate includes all agriculturally related activities, such as processing and marketing. These two activities, and small trading in non-food consumer items, are the major non-farm activities. Although the term non-farm is used to describe all post-harvest activities, some food processing is carried out on farms as a matter of convenience.

Nkwanta district has a very bad road system; even the main road from Jasikan through the district capital to Damako in the north is sometimes impassable during the rainy season.

Summary

The two study villages have a very small non-farm sector. The few non-farm activities are concentrated in the district capitals. In villages there are only one or two small traders with kiosks selling consumer items such as cigarettes, matches, soap and sugar. The situation urgently needs ways to create a vibrant non-farm sector, which reinforces the relevance of the present study. Both districts have relatively low population density and poor infrastructure, which have serious implications for the strength of linkages. Low population density limits the size of local markets and causes land-surplus agriculture in which per capita land availability is much higher than the threshold for transfer of labour from the farm to the non-farm sector. Poor infrastructure, especially roads, creates transport bottlenecks that will adversely affect forward linkage activities.

LINKAGES AND THE CASSAVA MARKET

This section examines the variety of linkages between cassava production and the rest of the economy. The cassava market is divided for this purpose into the various types of products made from the crop, which allows a clearer evaluation of the links between cassava and the rest of the economy and is a mechanism for comparing linkages across products. The information gathered for this analysis is based on interviews conducted in October and November 1997 in the cassava-producing regions of Atebubu and Nkwanta with informants including farmers, processors, traders and buying agents, transporters and officials of the Ministry of Agriculture and district assemblies, who were asked specific questions about cassava production and the market for cassava products. Discussion of cassava product markets is preceded here by some general information gathered from the interviews.

Cassava has only recently emerged as an economically important crop in the study districts. For a long time, yams were the staple crop in these regions as well as the major cash crop; cassava was intercropped with yams. In Atebubu, cassava was left for the women to harvest after the yams had been harvested. The economic status of cassava in the two districts began to change after the 1982–1983 drought and the ensuing famine. In the Nkwanta district before the drought and famine, ethnic groups in the southern part of the district processed cassava into *gari*, roasted granulated cassava. The ethnic groups in the north processed cassava into *kokonte*, dry cassava chips or flour with informants.

Cassava is primarily produced by small-scale farmers on fields of up to 0.8 ha. In Nkwanta district, smallholders constitute 70 percent to 80 percent of farmers. Farms with 6.0 ha and above are considered large-scale and account for only 5 percent of farmers. In Atebubu, average holdings are about 2 ha; two thirds are under cassava. Holdings of 8.0 ha and above are considered large-scale, accounting for about 10 percent of farmers. Cassava is normally intercropped with yams, maize or beans, but the dominant mixture is yams and cassava. There is little difference in cropping systems or technology levels between small-scale and large-scale farmers. Yam mounds are prepared from October to December; the yams are planted by February or March. Cassava is planted when the yams have taken. Yams are harvested after four to six months and the cassava is left to ripen. Depending on the variety, cassava can be harvested after 6 to 12 months. The very late varieties can take between 18 and 24 months to ripen. Ripe cassava can normally be stored in the ground for several months and harvested as needed, a feature that enables cassava to play an important role in household food security.

Farming in Nkwanta is based on minimal technology; hoes and cutlasses are the main farm implements. There is just one serviceable tractor in the district. Animal traction for ploughing and transport was introduced by the International Fund for Agricultural Development (IFAD) Smallholder Credit Input and

Marketing Project in 1994; as of November 1997, there were three sets of animal-traction technology working. Although tractor ploughing is more common in Atebubu, cassava fields are not ploughed because ploughing requires complete clearing of the land, which exposes cassava to the elements. Fertilizers are little used, and certainly not on cassava. It is claimed that soils in both districts are rich; crops do well without fertilizer. No pesticides are used on cassava and farmers are encouraged by extension personnel to use pest-free and disease-free planting material. Weeding is done by hand, although in Nkwanta it was claimed that one or two farmers use herbicides. The organization of cassava production thus presents few opportunities for backward linkages. Lack of demand for inputs such as mechanized equipment, fertilizer and pesticides means that little business activity can currently be generated in these areas to serve the cassava economy. Although hoes and cutlasses are in high demand in this low technology farming system, the demand generated for them cannot be attributed to cassava alone.

The most important purchased input in cassava production is hired labour for land preparation – clearing and mounding – weeding, harvesting and processing. The labour requirement for cassava production using traditional technology is estimated at 35 person-days/ha. The 17 000 ha cassava area in Atebubu creates about 600 000 person days of employment, about 75 percent of which is hired. Hired labour in Atebubu is mainly migrant labour from northern Ghana; in Nkwanta the hired labour comes mainly from Benin. Even for hired labour, therefore, the employment created does not directly benefit local inhabitants. It is not clear whether there is an indirect benefit to the local economy from the income earned by migrants, because seasonal migrant labourers return home with earnings for their own farming activities. The local economies would benefit if migrant labourers purchased farm implements or consumption goods while they were in the region.

Cassava processing offers the best opportunity for linkages from the farm sector to the non-farm sector. The description of the organization of cassava processing reveals the extent to which this potential for farm/non-farm linkages is being realized in the study area. Cassava products may be classified as traditional or modern. Traditional products, such as *gari*, *kokonte*, cassava dough and starch, are produced either for subsistence or for sale in local markets. Modern products, such as cassava chips, are produced for sale to processors for commercial purposes. The dominant traditional cassava products in the study area are *kokonte* in Atebubu and *gari* in Nkwanta. The major modern product in both districts is cassava chips for animal feed. Cassava chips are a new product, produced solely for one company. The different types of product markets and the linkages within those markets are now discussed.

Kokonte processing in Atebubu

Kokonte is only processed on-farm as an individual activity. Labour for processing is hired on a share-production basis whereby farmers hire labourers to uproot and process the crop; the product is then shared equally between farmers and labourers. There is little other productive activity generated from processing *kokonte*, because the technology is labour-intensive, based on knives and fuelwood. There is, however, considerable potential for employment generation locally because the labour hired for processing *kokonte* is local youth, mainly men, who form themselves into gangs. Some work gangs may include two or three women, who help to headload and peel the cassava. The young men who sell their labour for *kokonte* processing have small farms and therefore spare time to earn income from other sources. Women are involved more in the home processing of another type of *kokonte* called *chorchor*. The cassava is peeled at home, crushed in a mortar and dried in the sun. This type of processing is done on a very small scale at a rate of about one bag per week, equivalent to half a tractor load of smoke-dried *kokonte* for on-farm processing.

The marketing of *kokonte* involves a network of buying agents, handlers, transporters and traders. The role of each of these actors is set out below, but it is not easy to assess the number involved in each of these activities or the wages paid.

The buying agent is the link between farmer and trader, who is usually from a large town or city. Each agent has a number of regular traders for whom they identify sources of *kokonte*. With the help of handlers, agents provide bagging services for the trader. Handlers are hired on a casual basis; the number hired depends on the quantity of cassava to be bagged. When a *kokonte* buyer is identified, the farmer arranges for a tractor to transport the product from the farm to the house for bagging. The tractor driver is helped by workers; they assist the farm labourers, who processed the *kokonte*, to load the loose product onto the trailer. The charge for the tractor includes the services of the tractor workers. One load is between 40 and 50 bags; there may be up to ten trips per farm. The trader pays only after the product has been bagged. Traders operate in groups of five or six, and buy the product weekly or fortnightly. The duration of one buying trip depends on the availability of *kokonte*. It is understood that some traders are agents themselves, working for others in big towns or cities.

The next stage in the marketing chain is transporting the product out of the district. Depending on the volume, the trader may arrange for a truck for long-distance haulage, or another tractor to take the product to the market in the district capital to be transferred to a bigger truck. The trucks are served by handlers called loading boys. Weekly haulage of *kokonte* from the district market ranges between two and five trucks with capacities of 7–10 mt or between 120 and 180 bags. One trader is capable of purchasing one truckload of the product at a time.

As explained earlier, the proceeds from the *kokonte* are shared equally between the farmer and the gang of labourers. The value shared is the net return after deducting the cost of tractor services. The tractor and truck services and handlers are supplied through the local branch of the Ghana Private Road Transport Union (GPRTU), which sets prices for these services. Transport charges are paid by the bag and depend on the distance covered. The rates for tractor haulage varied considerably as stated by GPRTU informants and farmer and agent informants. The GPRTU informants quoted rates between 400 *Cedis* and 1 000 *Cedis* per bag; farmers and agents quoted rates between 1 500 *Cedis* and 3 000 *Cedis*. The charge for bagging the produce is 500 *Cedis* per bag; the fee for loading sacks of *kokonte* onto trucks is 200 *Cedis* per bag; both are paid by the trader, who also pays the cost of trucking out of the district.

There are often special financial arrangements between farmers, traders and agents. A trader may provide a loan of between 100 000 *Cedis* and 200 000 *Cedis* to a farmer against the supply of *kokonte*. The amount of a loan depends on the quantity of *kokonte* a farmer expects to process. The agent assures the buyer that the farmer has cassava for processing. Another type of financial arrangement is a farmer providing *kokonte* on credit to a trader, who can take delivery of the *kokonte* and pay on the next purchasing trip, which may be in a week or two. The maximum of such credit is usually 55 bags or a full tractor load. It is usually large-scale farmers who engage in these financial transactions. A large-scale farmer receiving a loan has a better chance of guaranteeing it with available cassava; a large-scale farmer is more likely to be able to supply cassava on credit as a loan. Farmers who can provide this kind of credit establish regular demand for their product and can protect themselves against falling prices. These financial arrangements are sufficient provided the scale of processing remains at the current level. Any increase in scale must come through adoption of improved processing technologies, which will require higher levels of financing that traders alone may not be able satisfy.

A number of farmers indicated that income from *kokonte* is used principally to pay for land preparation in yam and cassava fields. Farmers are thus able to increase production of yams, which is the main staple and cash crop. The first of two peaks for *kokonte* processing in November to March reflects farmers' need for cash to pay for land preparation. The second peak in June to August is a period when there is less farm activity.

Gari *processing in Nkwanta*

Gari processing is an individual income-generating activity that relies solely on family labour. Unlike *kokonte* processors, *gari* processors tend to buy cassava to add to their own production. Farmers sell cassava by fields of 0.4 ha. The price depends on the ripeness of the crop, but is linked to the price of a 50 kg bag. There are no buying agents in this trade. Although farmers with bigger fields sell cassava, farmers tend not to have a preference as to the size of processor they sell to. Farmers supply cassava to processors, who pay after the processed product has been sold. The longest repayment period is normally two weeks. The financial arrangements between farmers and processors are based on trust, developed through long-standing relationships. The cassava is purchased from local farms at distances of up to 10 km.

Gari processors do not hire labour, even for the harvesting. There is no labour supply, because people prefer to engage in production for themselves, no matter how small the scale. Processors call on relatives to assist them if they need more hands. One person working alone is able to process 50 kg of *gari* per week; with two workers, 50 kg can be processed in two days.

Grating is the only part of the process that is mechanized; a few processors use mechanical presses to remove the liquid from the grated cassava. Processors prefer mechanized grating because it is faster and produces a finer product. Manual processing of one unit of cassava field would take ten days; it can be grated in a day using a mechanical grater. Mechanical graters are normally owned by an individual who provides the grating service for a fee. Some farmers now use mobile powered graters, which permit them to process on farms. In such cases, the processors may hire workers to help with the roasting. This latest system of processing and the related linkages need to be investigated in more detail.

Although processors can sell to retailers and wholesalers, their main customers are traders from big towns and cities such as Accra, Ho and Nkwanta. There are also buyers from Burkina Faso and Niger. Processors also sell in markets at the district capital by carrying on their heads enough to meet current cash needs. The buying season for traders from outside Ghana is from January to July. Traders from towns in Ghana buy less during the rainy season, when driving on the main

TABLE 2
Rates of payment to actors in the *gari* trade

Actor	Payment per bag (*Cedis*)
Agent	500
Handlers who bag the product	500
Handlers who sew the bags	200
Transporters	500
Owners of storage space	200 (irrespective of duration)

Source: Author interviews.

road to the district becomes difficult. The price of *gari* varies according to the availability of other foods: prices are relatively high in July and August; they are very high in October to early December.

The system of *gari* marketing in Nkwanta also uses agents, who identify sources of the product and buy and package it for export out of the district. A trader advances money to the agent based on the current market price. Traders often advance money to processors to guarantee supplies. Price variations do not affect the financing contracts. There are sometimes barter arrangements whereby traders provide processors with non-farm consumer items such as clothes in return for *gari*. Conflicts can arise from these contracts when processors default. Agents act as intermediaries between the contracting parties, disbursing funds and ensuring that traders get the product within the specified period.

After buying *gari*, agents engage handlers to bag the product and sew the bags. Transporters move the product into temporary storage. Agents sometimes provide storage space; otherwise, there are property owners with spare rooms who provide the service. Payments to different actors in the processed cassava business are shown in Table 2. All payments are piece-rate and depend on the number of bags, irrespective of time taken. Agents prefer to deal with foreign traders from the West Africa subregion, because they make prompt payment for services; the larger volumes purchased by foreign traders create more business.

Chip processing in Atebubu

The TCG, the sole exporter of cassava chips, is responsible for organizing farmers to produce chips. About 4 000 farmers supply chips to the company; of these, 2 500 are considered regular and reliable suppliers who have taken up cassava chip production as a regular source of income. These farmers do not buy cassava to process; the company does not encourage farmers to buy cassava for chipping,

TABLE 3
Chip output by farmer category

Category	Seasonal output (52 kg bag)	Proportion of farmers (%)
Small-scale	1–5	20
Medium-scale	5–15	50
Large	15 and above	30

Source: TCG area office, Atebubu.

because it increases costs for the farmer and introduces a risk of irregular supplies to the company. The farmers include small-scale and large-scale farmers with farms of between 1.2 ha and 16 ha. In terms of chip production in a season, farmers can be grouped as shown in Table 3; some farmers in the large-scale category can chip up to 900 bags per season.

Chip production is an individual income-generating activity. Farmers may use family labour or hire labour for harvesting, but they usually do not hire labour for chipping; some large-scale farmers may hire labour for chipping on share terms similar to those for processing *kokonte*. There is no use of reciprocal labour in chipping; the chippers' association in Atebubu district does not engage in communal chip production.

Chips are processed manually and by machine. TCG has 31 chipping machines, although it is estimated that 90 percent of farmers do manual chipping. Most farmers prefer to chip by hand because the chips are easier to dry and hand chipping gives higher yields. The chipping machine tends to produce small chips, which require more careful drying and become powdery more easily. Only large-scale farmers use the chipping machines; more farmers opt for machine chipping at the beginning of the farming season, when demand for farm labour is at a peak. Processing *kokonte* is hard work, which is why farmers hire labour to do it. Processing chips is much easier and can be done casually by farmers in their free time. An estimated 40 percent of chip processors are women. Before the advent of the cassava chip market, men donated their cassava fields to women to process the crop into *kokonte*. Now that cassava has become valuable and cassava is fully integrated into the cropping system, men's attitudes have changed with regard to disposal of their crop; the result is that women themselves are growing it. There is no problem of access to land; the only problem that farmers face is the high cost of labour.

Farmers tend to grow only varieties with high dry-matter content for chipping. Most of the local cassava varieties ripen in a year, but there is one that ripens later and can remain unharvested for about three years. This variety gives farmers the flexibility of chipping at any time during the year. It was only during the 1997 cropping season that one of the varieties released by the Crops Research Institute was introduced for testing by a few farmers in the district.

TABLE 4

Rates charged for hauling chips

Distance range (km)	Rate charged per 52 kg bag (*Cedis*)
1–8	100
8–16	200
17–24	300
25–32	400
More than 32	500

Source: TCG Area Office, Atebubu.

The export company is the sole buyer of chips. The company organizes processing areas into buying districts, made up of zones and each zone has a buying centre. There are four buying centres, comprising 16 zones. In the Atebubu area, the cassava chip-buying district includes the administrative districts of Atebubu and Sena west.

Cassava chips are normally processed on-farm, as with *kokonte*. If rains set in when farmers are chipping, however, the cassava is transported home or sent to company warehouses for chipping. In such cases, the company provides transport at a cost to the farmer shown in Table 4. Although fresh cassava is transported, the rate charged depends on the volume of chips produced; farmers face the same transport cost whether they processes on-farm or off-farm. There is therefore an element of subsidy to the farmer who processes at home.

Linkages to the local economy for chipping are limited. Farmers hire labour for land preparation, mounding, weeding and harvesting of cassava. But much of this labour is seasonal migrant labour. TCG creates employment for local people as agents, loaders and tractor operators. Little business activity has emerged to serve farmers or processors. The main types of business activities are tractor hire and transport. The chip export company provides subsidized tractor and transport services. The company hires tractors and 7-ton trucks for hauling chips from farms to weighing centres.

Many farmers suggested they had increased expenditure on consumer durables and farm implements from cassava chip earnings. TCG realized this, and provides some of these items to farmers on credit against delivery of chips, including cutlasses, hoes, rubber boots, cement, roofing sheets, bicycles and iodized salt. Some farmers have requested television sets, radios and sewing machines.

Yam production is increasing, because the higher earnings from cassava are reinvested in bigger yam fields, which means interplanting with more cassava. About three chippers have purchased pickups to provide transport between Atebubu and Kumasi. Farmers are also investing more in children's education; many cited their ability to pay school fees and other education-related costs as a benefit of processing chips.

Before the introduction of chips, *kokonte* was the major cassava product in the region. Estimates from various informants suggest that between half and two thirds of annual cassava production is now processed into chips. Although *kokonte* is strictly for human consumption, cassava is sometimes sold to the chip market when the *kokonte* market is down or when farmers need cash in lump sums. The major attraction for chips is these lump sum payments. It appears that the chip boom is experiencing a lull, however, because cash-flow problems have limited the ability of the company to make timely payments. Maintaining farmers' interest in processing chips will depend on the ability of the company to avoid delays in payments.

It is generally believed that cassava production is increasing because of the alternative market for chips. Large-scale farmers switch between processing chips and *kokonte*, with the choice depending on:

- the processing preferences of hired labour, because hired labour is paid in kind;

- the weather, because it is easier to process *kokonte* in the rainy season;

- farmer's need for cash at the harvest time;

- distance between farms and villages, which determines transportation costs paid by the farmer; it is more expensive to transport *kokonte* from distant farms.

Although few farmers plant cassava only, the density of cassava in yam fields is increasing. Income from cassava is increasing because of increased output and better prices. The chip market has helped to boost *kokonte* prices, because chips and *kokonte* have become competing alternatives, putting pressure on *kokonte* traders to improve prices. During the 1996/97 season, for example, chips were bought at 4 500 *Cedis* per bag and *kokonte* at 12 000 *Cedis* per bag. By the 1997/98 season, the price of chips was 5 000 *Cedis* while *kokonte* was already selling for 20 000 *Cedis* per bag. This trend is another threat to the viability of the chip industry: farmers are beginning to ask for higher chip prices because of the relative increases in the price of *kokonte*. Traders in *kokonte* are also experiencing higher transaction costs for each purchasing trip, because it takes more days to organize a trip now than before farmers went into chip processing.

Cassava chip processing in Nkwanta district

Although the organization of the cassava-chip market in the two districts is similar in many respects, some significant differences arise from the alternative uses that farmers have for their cassava.

Participation in chip production in Nkwanta is much lower, estimated at about 400 core suppliers, most from the northern and eastern parts of the district. The south of the district is dedicated to *gari* processing, and farmers are still nervous about the chip market. As in Atebubu, cassava chipping is an individual activity, using cassava from people's own farms. About 15 percent of the major chip producers are women, and numbers are increasing. Processors may be classified as small-scale, producing from 1 kg to 500 kg per season, medium-scale, producing from 500 kg to 2 500 kg, and large-scale, producing over 2 500 kg.

Farmers are advised to harvest cassava at 9 to 12 months, because this creates the high dry-matter content. Technologies for chipping include manual and mechanized chipping. The company provides mechanized chipping services at a small fee, which is reflected in a lower price than for manually produced chips. In the 1996 season, for example, the cost of mechanized chipping was 12 percent of the price of 4 500 *Cedis* per 50 kg. As in Atebubu, farmers prefer hand chipping, but for different reasons. In Nkwanta, farmers assist each other with labour through the cassava chippers' associations. Labour used for chipping is family labour or communal/exchange labour. There is therefore no employment created for chip processing apart from the company's own labour force. The dominance of hand chipping also limits any direct production linkages. Chippers require no services that the non-farm sector can provide.

Chip processors tend to spend their earnings on roofing sheets, child education and bicycles. Many more people are building houses, although they are still made of mud. There is also conspicuous social spending on such events as lavish funerals at which abundant food and beverages are served.

The major problem facing the chip industry now is the company's lack of funds to purchase chips, especially during the peak season. A growing demand for *gari*, resulting from an influx of traders from the West African subregion, and a cultural preference for *gari* in the southern parts of the district present a challenge for the buying company in terms of improving its cash flow. *Gari* processing is increasing, and in 1997 the price was fairly stable. This trend is probably a result of the export marketing of *gari* that started sometime between 1996 and 1997 and caused the decline in chip production in Nkwanta district, as reported by TCG. Although farmers say that low chip prices are the disincentive, the difficulty of the work and the costs of processing chips and *gari* are not the same. Requirements for labour and other inputs for *gari* are high, but do not appear to worry farmers. Farmers nevertheless claim that processing chips is a better alternative when cassava begins to rot, because while *gari* processing can only be done in small amounts, a farmer can uproot a complete field of cassava and chip it all at once.

RURAL PRODUCERS, LINKAGES AND THE EVOLUTION OF THE CASSAVA MARKET

As noted in the previous section, the cassava market has evolved considerably in recent years. In this section, the prospects for expansion of cassava chips exports and the effect of changes in the market on farm/non-farm linkages are considered.

As in most of Africa, cassava utilization in Ghana is still primarily for human consumption, either as unprocessed roots or in the form of traditional cassava products – *gari*, dough and *kokonte*. The expected negative effect of urbanization on demand for some traditional cassava products can be overcome with adaptations of the products to suit the needs of urban consumers. The Food Research Institute has, for example, developed a cassava flour whose main advantage is reduced meal-preparation time. Cassava flour can be used as a wheat flour substitute for the baking industry (Day *et al.*, 1996). Even with these developments, external markets will be necessary if a substantial increase in cassava demand is to be achieved. In addition to exports of small quantities of traditional products, Ghana began to export cassava chips for livestock to Europe in 1994; earnings from chip exports recorded a dramatic increase from US$284 000 in 1994 to over US$2 million in 1997, followed by an equally dramatic fall to US$208 000 in 1998. The new export avenue for cassava presents opportunities for agro-industrial development, but access and responsiveness of rural producers to the new market opportunities are significant factors that set the pace of agro-industrial development. In the case of cassava, where the export product faces competition from processed traditional products, response to export demand will depend on the level of incentives derived from this market compared with those derived from processing for traditional markets. To understand these incentives, we need to investigate how rural producers fit into the cassava market.

Cassava farmers in the two districts prefer to process the crop themselves rather than sell the fresh root. In both districts, over 98 percent of fresh cassava processed by farmers is from their own farms; purchased fresh roots represent less than 20 percent of fresh cassava processed. These results are consistent with data from the Collaborative Study of Cassava in Africa (COSCA), which suggests that producers in remote areas such as the study districts process a high proportion of their produce because of the bulkiness of fresh root (Nweke, 1996). Farmers do not specialize in cassava processing because:

• they can transfer farm labour during the off-season to the non-farm sector, a strategy essential for overcoming problems of labour redundancy that characterizes seasonality in farm production;

• processing improves cassava income and stabilizes overall income because of improved shelf-life;

• farmers lack confidence in the reliability of cassava supplies from other farmers; this also occurs because of variations in supplies as farmers respond to annual price variations.

Because they do not specialize in processing, many small-scale processors widely dispersed across the producing areas use household labour and traditional technologies. The small size and dispersal of processors in turn require an efficient network to assemble adequate volumes.

The marketing chain for chips in Ghana is short, with the exporter dealing directly with the thousands of small-scale processors. In order to ensure assembly of exportable volumes of quality chips, exporters use an extensive network of agents and complicated logistics. To ensure stable supplies, exporters give farmers incentives such as credit for consumer items and subsidized transport between farm and local assembly depots. As the company reduces explicit cash payments to farmers, however, its own costs of assembly, transport to port and associated administrative costs account for 25 percent of cost of export brokerage (Table 5).

The scattered distribution of producers and the poor rural road networks account for the high transport costs. It is estimated that while production costs in cassava producing areas of Brong Ahafo are about 48 percent lower than costs in Greater Accra Region, the cost of transporting chips from Brong Ahafo to the port of Tema is 80 percent more than the costs to transport chips from Greater Accra to Tema (TechnoServe Ghana, 1994). Dadson, Kwadzo and Baah (1994) have also shown that in general, the competitiveness enjoyed at the farm gate by root and tuber crops is lost at the level of urban market because of high transportation costs.

TABLE 5

Component costs per ton, TCG (1998)

Item	Value (US$)
Price of produce (paid to farmer)	45.00
Transport, village buying centre to district depot	2.90
District depot to Tema	6.00
Operation of village buying centre	1.28
Operation of district evacuation depot	3.40
Packaging and handling	1.99
Headquarters costs	2.66
Financing (short term @ 15%)	9.45
Total Cost	72.68
FOB[1]	86.00
Exporter's margin on FOB (%)	18.00

[1] FOB: free on board.
Source: TCG area office, Accra.

Freight charges from Ghana are not globally competitive, ranging between US$35/mt and US$40/mt, compared with only US$9/mt from Thailand (David Pessey, TCG, personal communication, 1996). The high cost of freight has been attributed to inefficient handling at the port in Ghana. The loading rate at the port of Tema is 600 mt per day, compared with 20 000 mt achieved by the competition. The FOB price received by Ghanaian exporters is therefore far below the average world price; the difference was US$47/mt in 1996 and US$11/ mt in 1997. In addition to a volatile average world price, the price received for Ghana's chips has declined steadily from an average of US$120/mt between 1993 and 1994 to US$86/mt in 1997. The effect of low FOB prices and high costs of haulage are a price squeeze on producers, which affects the relative profitability of processing chips for exports.

Returns from processing various cassava products are used as an indicator of the relative incentive to process each product. The budgets for cassava production and processing of *gari, kokonte* and chips are estimated under two scenarios: i) farmers process their own cassava with hired labour, and ii) farmers process their own cassava with family labour. The latter scenario is based on the fact that farmers in the two districts generally process their own cassava without hired labour. Since farmers process their own cassava, fresh cassava is valued at the variable cost of production. Tables 6 and 7 show the results of these scenarios.

TABLE 6
Scenario 1: Farmers process their own cassava using hired labour

Indicator	*Gari*	*Kokonte*	*Chips*
Net earnings per mt (*Cedis*)	121 567	39 835	31 255
Net earnings as % of total cost	32	23	60
Returns to hired labour (*Cedis*)	2 380	7 620	1 190
Total labour input per mt of product person-days	70	9	20

Source: Author's calculations.

Net earnings from cassava processing are estimated at 31 000 *Cedis* for chips, giving approximately 60 percent return on investment, 40 000 *Cedis* for *kokonte*, giving 23 percent, and 122 000 *Cedis* for *gari*, giving 32 percent (Table 6). There are no previous estimates of returns on investment for *kokonte* or chip processing. The estimate of 32 percent for *gari* is higher than the 18 percent estimated by Agyako-Mensah (1985) but comparable to the 27 percent margin for Kumasi processors in Kreamer (1986). The difference in results is because of the valuation of fresh cassava in this study at variable cost rather than at the market price. The higher performance of chips over *gari* and *kokonte* is because of the lower cost of hired labour, which for all three products is costed at a third of the value of the final product.

The second scenario of farmers using their own labour is to demonstrate the returns to family labour engaged in processing (Table 7). These returns are uniformly higher than the returns using hired labour for all three products. The exceptionally high percentage margins, however, only illustrate the high family-labour component of total costs. The use of family labour is common in *gari* processing in Nkwanta, where earnings per mt of product are 1.2 times higher when family labour is used than when labour is hired. The difference between payments to hired labour and net returns to family labour is 41 percent in Atebubu, which is probably not high enough to entice family labour into the tedious task of *kokonte* processing; hence the widespread use of hired labour for processing *kokonte*.

The budget estimates demonstrate the low level of incentive for processing chips compared with *gari* and *kokonte*. Net earnings per day of work by family labour are lowest for chips. Farmers complained that the price of chips is too low and noted that the only reason they process chips was because of the possibility of receiving payment for their product in bulk, a payment structure which allows farmers to process chips to meet specific cash needs. Farmers

TABLE 7
Scenario 2: Farmers process their own cassava using family labour

Indicator	*Gari*	*Kokonte*	Chips
Net earnings per ton (*Cedis*)	268 233	55 000	59 030
Net earnings per day (*Cedis*)	3 830	7 700	2 950
Net earnings as % of total cost	115	109	243
Total labour input per ton of product	70	9	20

Source: Author's calculations.

admit, however, that the good prices they receive for the traditional products are a result of the introduction of chips.

The structure and relative profitability of chip production and the market trends lead to identification of the following challenges for the expansion of chip exports, categorized as farm level, national and international constraints:

• **Farm level.** The price of chips and the profitability of chip production are lower than those of traditional cassava products, which have the added attraction of being food staples for farmers. Farmers are not familiar with the use of chips as animal feed. Small producers scattered over a wide area with poor roads and limited transport increase the costs of primary haulage. Although the use of a chipping machine can reduce labour input and contribute to production of chips of a standard size, it has not been taken up by farmers, who claim that the machine tends to slice the cassava too thinly, making the drying process more tedious. The machine tends to mash the cassava, thereby reducing potential yield of chips. These are technical problems, which ought to be addressed to make the technology more appealing. More important underlying reasons for the unattractiveness of machine chipping, however, are the small volumes produced and the fact that chip processing occurs during the slack labour period and poses no competition with farming activities.

• **National level.** High freight charges reduce the competitiveness of Ghana's chip exports. High interest on borrowing, about 30 percent in 1998, limits capital mobilization. During fieldwork in 1997, for example, it was clear that cash-flow limitations on export brokers had caused frustrating delays in paying farmers. Insufficient domestic demand for chips is another factor limiting expansion of the market. Despite a wide range of local research on the use of cassava as an energy source in the feeding of poultry, pigs and small ruminants, maize remains the main source of energy in feed compounds.

None of the 27 members of the Ghana Feed-Millers Association uses cassava as an ingredient. The reasons include lack of knowledge on the use of cassava as feed, additional costs to be incurred in adapting mills to cope with the larger bulk volume and dust emissions from milling cassava chips. With an annual output of only 400 mt per annum, the feed-mill industry cannot have a substantial impact on cassava demand. Finally, cassava utilization is highly dependent on the availability of other major staples: the higher the level of food availability, the lower the demand for cassava. Levels and stability of export supplies will therefore depend on the levels and stability of preferred staples, especially maize.

- **International level.** The cassava chip market has been in decline over the last decade. World exports declined from 11.9 million mt in 1989 to 3.3 million mt in 1997, driven largely by a decline in European imports. The Netherlands reduced imports from 2.5 million mt in 1980 to 1.6 million mt in 1989, and to about 1 million mt in 1997. Germany's imports declined from 1.3 million mt in 1989 to about 148 000 mt in 1997. The decline in demand for cassava chips in Europe has been attributed to the level of intervention prices for feed grains in the European Union, which favours the use of domestic feed grains (Henry and Gottret, 1995). Ghana's entry into the international chip market can be described as untimely, coming at a moment when the market was declining. As of 1996, the country was hoping to access a General Agreement on Tariffs and Trade (GATT) quota of 140 000 mt, but the will to participate in this market has declined.

The current state of the cassava chip market and the problems with Ghana's international competitiveness in this market have the potential to alter the cassava market radically. Even now, there is a strong incentive for many farmers to produce traditional processed products instead of chips. If the export market continues to falter, this trend is likely to continue. Collapse of the export market could lead to a dramatic reduction in demand for cassava and a reduction in the price for all processed cassava products. This will limit the beneficial linkages between cassava production and processing and the employment market, erode incomes and reduce expenditure linkages.

INSTITUTIONAL SUPPORT AND THE CASSAVA SUBSECTOR

Institutional support for the cassava subsector comes from a number of sources, including the public sector, research centres and the private sector. Support to the sector includes identification of markets, development of new products and development and promotion of processing equipment. In this section, the support received by the sector and ways in which it could be improved are considered.

The Government's declaration of 1995 as the Year of Cassava reflects growing interest in the crop at official levels. Following this, a 15-member multidisciplinary Cassava Task Force was set up in March 1996 by the Ministry of Food and Agriculture to oversee development of the crop. The group's terms of reference included identification of appropriate research for improving production and processing, and promotion of the crop to local and foreign investors. At a workshop on cassava processing held in September 1997, it was reported that US$11 million was being made available by IFAD to support development of roots and tubers, including cassava. It is expected that much of the funding will be devoted to multiplication of improved planting material. This support follows similar funding provided by IFAD and the Government since the late 1980s through the Root and Tuber Development Programme.

The Crops Research Institute released three cassava varieties in 1990, which have had varying degrees of success across the country. Adoption of the varieties has been more successful in areas where cassava is processed, for example the Volta region, than in areas where cassava roots are cooked without processing, such as the Ashanti region. The link between productivity increases and adoption of improved production technologies on the one hand and strengthening of demand for processing on the other have been reported in southern Brazil and Thailand, where growing demand for roots to process into starch induced farmers to adopt new production technologies (Henry and Gottret, 1995). Although these links are more downstream than upstream, it is expected that in areas where improved varieties are adopted, crop management practices may have to change if the yield potential of the new varieties is to be realized. These changes in crop management practices may bring about the necessary linkage activities in the non-farm sector.

It appears that the rate of spread of new varieties is rather slow. Trials for the introduction of new cassava varieties in the two districts of this study began only in the 1997 cropping season. These are both major cassava-producing districts, and it could be expected that such areas would be among the first to receive new production technologies.

The Food Resources Institute has developed fortified *gari* as a convenience food, and dry cassava flours as substitutes for bulky fresh roots and wet pastes. These products are primarily targeted to urban consumers. The products have not, however, gone beyond exhibition at fairs such as the Industry and Technology

Fair. The reasons why businesses have not taken up these products should provide leads on the appropriate strategies for product development and promotion. The experience of the Food Resources Institute with uptake of these products by the private sector suggests that individuals are only interested in the distribution aspect, especially export as ethnic food, but not in the processing aspect. The major constraint has been lack of financing to acquire the expensive processing equipment needed for commercial processing. The Food Resources Institute has also identified the absence of an industrial extension as a limitation on adoption of new processed products by the private sector.

The Natural Resources Institute has developed a process for storing fresh cassava, which has a good market in urban areas. They have also studied the market for *kokonte* and are following this up with technologies to improve the quality of the product and assessment of consumer preferences and the implications of the technology on costs.

Another important development initiative from the Government, which will benefit cassava production and marketing, is the Village Infrastructure Project. An important component of the project is development of village-to-farm tracks, to be complemented by the introduction of intermediate means of transport to replace carrying loads on the head. Such a development can facilitate movement of cassava and help to strengthen the linkages between cassava farming and the non-farm sector.

Recent private-sector entrants into the cassava industry are Glucoset Ghana Ltd, which has established a cassava plantation for processing starch, and TCG , the sole exporter of cassava chips as animal feed to the European Union. TCG has expanded operations from Nkwanta district in the Volta region to other parts of the country, including the Brong Ahafo, Ashanti and Northern regions. Exports have expanded from 30 mt in 1993 to more than 20 000 mt in 1996. The company is developing a partnership with farmers through incentives such as credit for consumer items and subsidized transport as a way of sustaining farmers' interest in chip processing. As the company tries to cut down on cash payments for transportation by farmers, however, the high cost of transporting chips from producing centres to ports is one of its major constraints. Regions with a comparative advantage in the production of cassava chips lose the advantage by the time the product reaches the port, because transport costs are so high.

The Sasakawa Africa Foundation is disseminating cassava-processing equipment, especially to women's groups. The equipment includes powered and manual equipment for processing chips and *gari*. The chipping machine, however, produces chips that are too thin for the export market. The *gari* processing equipment is more promising; it is now being manufactured by the Intermediate Technology Transfer Unit (ITTU) in Ho, the Volta regional capital. The portable set of equipment, which includes a grater, press, bagging stand, fermentation rack and sifter, is driven by a petrol engine. Capital requirements are high, however: unit cost of the machine as of November 1997 was 2.5 million *Cedis*, with a 700 000 *Cedi* import component on the engine. Apart from the engine, which is imported, all parts of the machine can be serviced by any technician. The machine can grate 1.2 mt of cassava per hour and has a fuel consumption of 4.5 litres per mt. It requires two people to operate; it can be assembled in 12 days by three people working approximately six hours a day. So far, three sets of equipment have been manufactured; two have been sold. ITTU aims to interest manufacturers of food processing equipment to take up commercial production of the equipment through training workshops.

Lakai Motor Company Ltd. developed a cassava chipper with the capacity to chip 46 mt of cassava per day. Initial sales were made to TCG, who provided services to farmers for a fee. Uptake of the chipping machine has been very slow; this was later revealed in the field to be because of lack of patronage by farmers.

CONCLUSIONS

Based on the information presented in this chapter, the following conclusions can be drawn:

1. Cassava production in Africa is not backward-linkage friendly. The technology applied in production does not generate demand for inputs other than labour. The employment created for fieldwork has limited linkage benefits for the local economy because of repatriation of earnings by migrant labour. The type of technology applied in cassava production is consistent with the requirements of the crop. Mechanized land preparation exposes the crop to the elements; fertilizers are not used because the soils are fertile. Cassava is noted for its adaptability to a wide range of growing conditions. Farmers are encouraged by extension agents to use clean planting material to prevent disease and pest infestation rather than to control them with chemicals. The conclusion is that promotion of low external input use as a strategy for sustainable agricultural production limits the potential of the farm sector to link up with the non-farm sector.

2. Forward production linkages from the farm begin with transport of produce to markets or processing centres. In Atebubu, marketing of fresh cassava is virtually non-existent. This is a deliberate strategy to reduce the bulk of the produce and make it more easily transportable. The poor condition of roads to production areas limits transport to a small number of tractors. In Nkwanta, where fresh cassava is sold to processors, the strategy is for farmers to shift the burden of transport to the processor by selling fields of ripe cassava instead of harvested produce. The processors harvest small amounts at a time according to batch size. The small amounts harvested are easily transported by trucks or headloaded to the processing centre, which is usually the processor's home. The reported use of mobile powered graters on farms is an even more advanced response to the poor road conditions. Poor road infrastructure in the districts therefore limits forward linkage activities in transporting fresh cassava, and supports the need for infrastructure development as a prerequisite for rural growth.

3. One traditional and one non-traditional processed product are identified in each of the districts. Processing of both products generates employment for local residents, particularly during slack periods in the cropping season. *Gari* processing has direct production linkages, in the form of grating services. The size of this service sector is limited, however, by the scale of processing undertaken. Reliance on family labour for such a labour-intensive activity inevitably reduces the scale of processing.

4. Marketing of chips and traditional products generates additional employment primarily through handling and transport. Productive activities related to packaging are non-existent, because packaging materials are imported. The relative earnings of local agents and handlers in the marketing chain are probably much lower than earnings of traders and companies. The major inputs for transport – vehicles, fuel and spare parts – are imports; at best they are supplied from the major towns in the district. All these situations represent leakages from the local economy.

5. The propensity to spend on consumer and investment items from cassava earnings appears to be strong, especially when farmers receive bulk payments for sales of processed products. Investment expenditure on employing more labour to expand farms promotes growth of the farm sector; expenditure on children's education promotes the development of human resources. Indications of consumption expenditure items, however, suggest that consumption preferences favour imports and may not be conducive to local growth.

6. Introduction of the chip market into the two districts and the export market for *gari* in Nkwanta have boosted prices of cassava products, and probably real farm incomes. The response to these new opportunities will determine whether the linkages between the farm and the rest of the rural economy will be strengthened. The response will depend largely on the level of support for local economies in terms of improved roads. With respect to labour-saving techniques, detailed assessment of farmers' needs and preferences should be undertaken to guide any further research into technology development.

7. The chip market has increased demand for cassava and boosted the price of all cassava products, but weaknesses in the market for cassava chips, partially because of local conditions in Ghana, have the potential to reduce the importance of this market and thus reduce demand for cassava products. Such an occurrence could hurt farmers and those linked to the cassava market. To avoid this outcome, the government must consider actions such as improving the transport infrastructure and reducing freight costs, which would assist the competitiveness of Ghana's cassava subsector .

8. The public and private sectors are giving increasing attention to the cassava subsector. Government programmes aim to improve productivity and production, but private-sector initiative is expanding demand sources. These strategies can complement each other if the bottlenecks resulting from poor roads are improved and processors and exporters gain better access to funding sources.

REFERENCES

Agyako-Mensah, N.A. 1985. *The impact of small-scale gari processing industry on agricultural production in the Ashanti region.* Kumasi, University of Science and Technology. (Unpublished thesis)

Dadson, J.A., Kwadzo, G.T-M. & Baah, K.O. 1994. *Structural adjustment and marketing of roots and tubers in Ghana.* Accra, FAO. (Report for FAO Regional Office for Africa.)

Day, G., Graffham, A.J., Ababio, J. & Amoako, M. 1996. *Feasibility study: market potential for cassava-based flours and starch in Ghana.* Accra, University of Ghana, Department of Food Science and Nutrition; Greenwich, London, Natural Resources Institute.

Henry, G. & Gottret, M.V. 1995. Cassava technology adoption: constraints and opportunities. In *Cassava breeding, agronomy research and technology transfer in*

Asia: proceedings of the fourth regional workshop held in Trivandrum, Kerala, India, 2–6 November 1993, pp. 410–432. Bangkok, CIAT.

Jebuni, C., Asuming-Brempong, S. & Fosu, K.Y. 1990. *The impact of economic recovery programmes on agriculture in Ghana*. Accra, USAID. (Report.)

Kreamer, R.G. 1986. *Gari processing in Ghana: a study of entrepreneurship and technical change in tropical Africa*. Ithaca, NY, USA, Cornell University Department of Agricultural Economics. (Cornell International Agricultural Economics Study.)

Ministry of Agriculture. 1990. *Medium-term agricultural development programme.* Accra.

Ministry of Food and Agriculture. 1995. *Proceedings of workshop on cassava as a substitute in the feeding of poultry and livestock, Kumasi, 11–12 September 1995.* Accra.

Ministry of Food and Agriculture. 1996. *Agriculture in Ghana: facts and figures.* Accra.

Nweke, F.I. 1996. *Cassava: a cash crop in Africa.* Ibadan, Nigeria, International Institute of Tropical Agriculture. (COSCA Working Paper No. 4.)

Seini, A.W., Nyanteng, V.K. & van den Boom, G.J.M. 1997. Income and expenditure profiles and poverty in Ghana. *In* W.K. Asenso-Okyere, G. Benneh & W. Tims, eds., *Sustainable food security in West Africa*, pp. 55–78. Boston, Kluwer Academic Publishers.

Stryker, J. & Dumenu, E. 1986. *A comparative study of the political economy of agricultural pricing policies: the case of Ghana.* Washington DC, World Bank (unpublished).

TechnoServe Ghana. 1994. *Feasibility report on cassava chips enterprise in Ghana.* Accra.

World Bank. 1993. *Ghana 2000 and beyond: setting the stage for accelerated growth and poverty reduction.* Washington DC, Africa Regional Office, Western Africa Department.